SCIENCE A ROAD TO WISDOM

EVERT W. BETH

SCIENCE A ROAD
TO WISDOM

Collected Philosophical Studies

D. REIDEL PUBLISHING COMPANY / DORDRECHT-HOLLAND

DOOR WETENSCHAP TOT WIJSHEID. VERZAMELDE WIJSGERIGE STUDIËN

First published in 1964 by Van Gorcum & Comp. N.V., Assen, Holland

Translated from the Dutch by Peter Wesly

ISBN-13:978-94-011-7646-0   e-ISBN-13:978-94-011-7644-6
DOI: 10.1007/978-94-011-7644-6

TRANSLATOR'S NOTE

References to Dutch scholars and publications have sometimes been omitted when this did not affect the continuity of the argument. Dutch titles have been omitted in both parts of the Bibliography unless they are referred to in the translation. The book by Halmos replaces a Dutch introduction to set theory.

I am greatly indebted to Miss Véronique Kroon for expert assistance in improving the English of the translation. The remaining imperfections are probably the cases where I did not follow her advice.

CONTENTS

FOREWORD

A few days before his death my husband requested me to write a few words of thanks on the publication of this collection of articles. He had already prepared the greater part of the volume for the press and had also decided on the title *Science a Road to Wisdom*. His original selection was somewhat more comprehensive, which is still partly reflected in the Preface. Knowing how much he wished to see this collection published, I respectfully and lovingly fulfil his request, thanking Else M. Barth and J. J. A. Mooij for their extensive and expert care in putting the final touches to the volume.

ADDITION TO THE TRANSLATION

Finally, I wish to thank Peter G. E. Wesly for his willingness to undertake the translation of the book into English.

C. P. C. BETH-PASTOOR

PREFACE

In this republication of a number of philosophical studies I have refrained from including articles of a specialized nature on symbolic logic and the methodology of the exact sciences. There was no cause to include my contributions towards the didactics of mathematics and physics, nor did I consider it appropriate to reprint pieces of a predominantly polemical nature. I decided, however, that a very modest selection from my purely historical work would not be out of place.

Of the four stages in my intellectual development as set forth in Essay XI, the first stage (up to 1935) is not represented, the second (1935–42) is represented by Essay VI, the third (1942–50) by Essays I, III and IV. The remaining essays (II, V and VII–XII) are characteristic of the position I have taken up since about 1950. Thus the earlier periods are represented comparatively scantily, the explanation being that I only started publishing towards the end of the first stage and that most of these publications are not suitable for this collection, for the reasons stated above.

Two essays (IV and V) originally appeared in English, the others in Dutch. Of the essay which is now VIII, the middle part has been dropped. I have made corrections in most of the papers, as well as some small modifications ensuring the necessary uniformity of style, spelling, notation and, as much as possible, of terminology. Moreover, I have supplemented the bibliographical references and combined them into a single bibliography.

Although I should hardly wish to answer for Bergson's well-known dictum *Un philosophe digne de ce nom n'a jamais dit qu'une seule chose* (nobody deserving the name of philosopher has ever said more than one single thing), it still reassured me when I found that in some cases the same line of thought, with slight variations, returned in several articles. If all the alternative versions but one had been dropped, stylistically unacceptable gaps would have appeared. Indeed, a number of alternative readings may facilitate an accurate comprehension, while the recurrence of the same line of thought in several different contexts gives an impression of its fruitfulness. Taking all this into account, I have decided on the following policy. Of each set of variants, one has been rendered as completely as possible; the remaining variants have been shortened as much as the context allows and have been provided with a reference to the complete version.

The material has been arranged systematically, not chronologically, for the aim of this volume is, above all, to give the reader a coherent and more or less complete survey of my ideas and my work in the field of general philosophy, so that, I hope, there will be less misunderstanding and uncertainty about it in the future. It is probably clear now that I have not aimed at being exhaustive. For further elaboration and more detailed explanations I refer to my monographs.

This book is intended for all who are interested in philosophy. A reasonable amount of general knowledge is required, and the reader should not object to some intellectual labour: a book such as this is not meant as light entertainment. I should be particularly gratified if it were to be read by the younger generation.

It cannot be denied that many young people, especially those destined by their talents and education to occupy leading positions in our society in the near future, are much given to irrationalistic trends. This is quite understandable; it is explained by the decline of traditional rationalism and by the strongly increased attention for and knowledge of the irrational forces in man. Moreover, ours is a time of unprecedented and yet ever-increasing rationalization of all forms of human life, and seeking refuge in irrationality is no doubt partly a natural reaction to these conditions of our time.

One does not, however, have to be an extreme rationalist to realize that unbridled irrationalism can hardly be conducive to the adjustment of the intellect to modern man's present and apparently self-chosen environment, or to be concerned about the eventual consequences of the lack of adjustment. If we wish to enjoy the benefits that our rational civilization has to offer without understanding the foundations on which it rests, we shall not be able to make a critical evaluation of undesirable forms of rationalization or to recognize dangerous developments before it is too late. We would thus be powerless against the collectivism that could proceed from undesirable forms of rationalism combined with certain manifestations of the irrational drives in man.

In order to adjust the intellect to contemporary conditions of life, to make a critical evaluation of the different forms of rationalization and to control the irrational forces in man, we shall require scientific discernment and a capacity for rational discussion. Certainly, these requirements, while enabling us to act efficiently, do not guarantee sensible behaviour; nevertheless, the wisdom we are seeking should be closely connected with scientific insight.

I do not claim to have even so much as drawn the outlines of the modern rationalism which in my opinion our time stands in need of, though I believe to have done some preliminary work for the development of such a rationalism. In the first place, I have pointed out some weak spots in the foundations of traditional systematic philosophy, which is predominantly rationalistic. These foundations date back mainly to Aristotle; my criticism does not, however, detract from my admiration for this thinker whose work has determined the structure and development of western philosophy for over 2000 years.

In the second place I believe to have given modern logic a form which renders it a suitable foundation for the development of a new epistemology and a new ontology. Moreover, the conditions have thus partly been created for the development of a new ethics which does justice to modern psychological insight and at the same time avoids the extremism of present-day irrationalism.

SCIENCE AS A CULTURAL FACTOR

Nowadays a widespread pessimism exists regarding the future of western civilization. This pessimism has been roused or, as the case may be, increased by
(1) cultural fatalism,
(2) the alleged foundational crisis,
(3) the social and political phenomena of our time.
In this essay I wish to inquire what there is to say about these three points from the – naturally restricted – view-point of exact science. I should like to clarify my argument by saying beforehand that from this point of view the prevalent pessimism – mainly influential in Western Europe, while it is practically unknown in the English-speaking world, and especially in Russia – cannot be considered as justified.

As I said, this view-point is restricted and so the value of my conclusion is limited, but it would be wrong to underestimate its significance. Firstly, the exact sciences exert a strong and ever-increasing influence on social and political life; and secondly, the very nature of the exact sciences allows a more reliable evaluation of results and possible future achievements than is possible in other domains of civilization.

Cultural fatalism, as it finds expression in the writings of Frobenius, Lamprecht, Pareto and Spengler, may be summarized as follows. Every civilization is distinguished by a certain system of values. Naturally the relevance of this system of values is limited to the products of a particular civilization and has otherwise no significance. The system of values determines the character and development of the civilization, which can be seen as a self-contained organism; after a process of growth, which takes an identical course in every civilization, it is destined to die out without leaving anything behind that could be of vital importance for the development of other civilizations. In this point of view there can, of course, hardly be any question of cultural progress.

By a comparative investigation of the rise and fall of various civilizations (historical 'morphology') the authors mentioned above arrive at the conclusion that western civilization is already past its bloom and is now in the last stage of its history, inevitably ending with its downfall.

Before proceeding to consider whether cultural fatalism is well-founded, I wish to emphasize its disastrous influence. This fatalism, together with the pessimism resulting from the foundational crisis and the social and political vandalism of our time add up to a vicious circle which, if allowed to continue unchecked, could seriously disturb the development of western civilization in the long run. We shall have to investigate whether it is possible to break through this vicious circle.

From the view-point of the exact sciences, cultural fatalism has been subjected to devastating criticism. Spengler (1918–22) had also applied his historico-morphological method to mathematics. According to him, every civilization has its own mathematics. Greek mathematics, for instance, deviates in essential aspects from modern western mathematics; the basic idea is different, and accordingly Greek mathematics has a character of its own.

This view has been strongly criticized by Scholz (1921) and by E. J. Dijksterhuis (1934). Historical data actually provide a very different picture.[1] The mathematicians and astronomers of the 16th and 17th centuries simply took Greek tradition as their starting-point; many of them were philologists no less than mathematicians. An example is Francesco Maurolico[2], who in connection with a proof by Euclid developed the principle of mathematical induction. This principle was applied by Pascal[3] when he introduced the famous 'triangle' named after him; it has played an important part in mathematics ever since. The peculiar method of reasoning based on this principle – a method which, as appears from the above, was already used by Euclid – was considered by Poincaré (1902) to be "le raisonnement mathématique par excellence". As far as this essential point is concerned there is a striking continuity between Greek and modern West European mathematics.

A similar picture of a steady development of mathematics, hardly influenced by the diversity of civilizations, can also be found in far earlier times. As Freudenthal (1946) has shown, the study of mathematics, in so far as we can trace it in the course of history, has always manifested the strongly international character which it still shows today. I should like to add that I am inclined to take a similar view as regards the study of philosophy (Beth, 1946/47, 1952/53).

It is not improbable that cultural fatalism has resulted from a one-sided orientation towards isolated and somewhat rigid civilizations, for example, those of Egypt and China. This isolation is, however, by no means typical of all civilizations; wherever it occurs it can be accounted for by quite distinct and easily identifiable causes. Besides closed societies, which the fatalists are inclined to concentrate upon and where their conclusions may hold good to a certain extent, there are 'open' societies; Greek culture is a characteristic example.

The conclusions of the fatalists as regards the future of western civilization are, in my opinion, ill-founded. Of course, I do not wish to deny that we are in a dangerous predicament in many respects; however, I think it extremely ill-advised to surrender to the idea of inescapable doom.

The second point I mentioned, the *foundational crisis*, must be dealt with in a more sophisticated manner. In the first place I hope to show that a foundational crisis actually exists in the field of the exact sciences as well. Then I propose to discuss the extent to which these critical phenomena justify the prevalent pessimism.

I do not want to embark upon enumerating all the time-honoured certainties that are called in question or even completely repudiated by modern science. I prefer to discuss a single case in more detail. This example will, I hope, make it clear that not rashness or presumption but experience and judicious reflection have lead to a break with established tradition.

Ever since the 17th century there have been two competing theories about the nature of light: the *undular theory*, defended principally by Christiaen Huygens (1629–95), who considered light as a process of waves, and the *corpuscular theory*, developed mainly by Isaac Newton (1642–1727), who explained optical phenomena by the motions of small particles. At first the corpuscular theory triumphed, but round 1800 Young and Fresnel published experimental results which could be explained only by the undular theory. So now the wave theory was triumphant, and in the course of the 19th century it was confirmed in a most brilliant way by experimental research.

3

After 1900, however, the theoretical work of Planck – following experiments by Lummer and Pringsheim and by Rubens and Kurlbaum – showed that certain experimental results can only be explained by the corpuscular theory (called *quantum theory* in this connection). In order to be able to account for *all* the experimental evidence it now became necessary to utilize in turn two incompatible theories – a very awkward predicament!

Using the work done by Heisenberg, Bohr (1929) provided the solution to this puzzle by means of an analysis, the essentials of which I propose to render schematically.

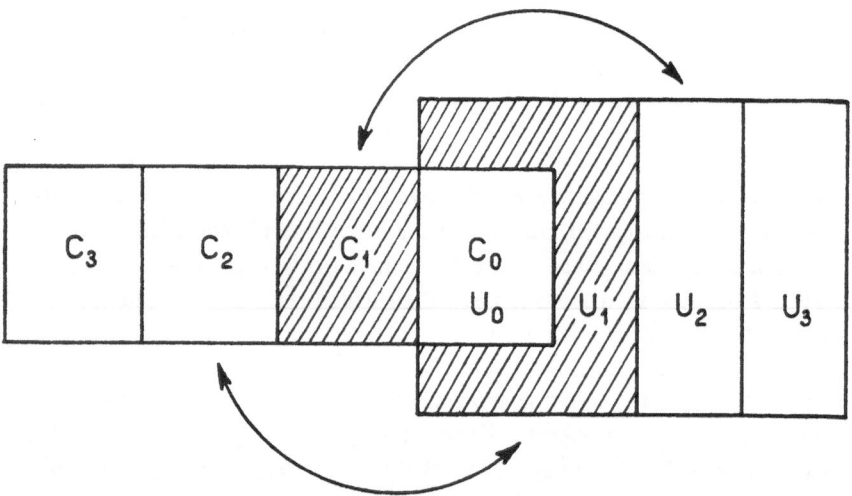

The inferences from the corpuscular theory are represented by the points of rectangle C, those from the wave theory by the points of rectangle U. Both rectangles consist of four parts: C_0, C_1, C_2, C_3 and U_0, U_1, U_2, U_3, respectively.

The areas C_0 and U_0 are co-extensive and represent the inferences that both theories have in common. These inferences are invariably confirmed by experiment. As an example I mention the laws on the reflection and refraction of light-rays.

The areas C_1 and U_2 represent the mutually contradictory inferences from the corpuscle and the wave theory, in so far as experiment shows

4

the wave theory to be correct. For example, the diffraction of light-waves, denied by the corpuscular theory, is predicted by the undular theory and is confirmed by experiment.

The areas C_2 and U_1 represent the contradictory inferences from the corpuscle and the wave theory, in so far as experiment shows the corpuscular theory to be correct. The above-mentioned experiments by Lummer and Pringsheim afford an example.

The area C_3 represents those inferences from the corpuscular theory which are neither confirmed nor contradicted by the wave theory; likewise U_3 represents those inferences from the undular theory which are neither confirmed nor contradicted by the corpuscular theory. These inferences are all confirmed by experiment.

Evidently, the difficulty lies in the occurrence of the areas C_1 and U_1. If we could eliminate the inferences concerned from both theories, they would no longer contradict but harmoniously supplement each other. Together they would supply an all-round explanation of the hitherto observed phenomena falling within the domain of the theory of light. There is, incidentally, a quite analogous situation in the theory of matter.

The principle of complementarity, formulated by Bohr, eliminated these embarrassing inferences from both theories with a single stroke. On closer investigation von Neumann (1932) and others concluded that acceptance of the complementarity principle amounts to a *weakening of logic*. For the elimination of the embarrassing inferences represented by areas C_1 and U_1 can be brought about, not by dropping one or more principles peculiar to the theories concerned, but by restricting the possibility of making inferences from these principles.

This is a veritable 'foundational crisis'. Until recently, a radical change in the established laws of logic, not by virtue of any aprioristic insight, but as a consequence of certain results of experimental research, was the very last thing that was considered possible. So it could have been expected that many would protest against this development in physics. Humanly speaking, these protests are understandable, but they lack objective foundation. As will be clear from the schematic explanation, acceptance of the complementarity principle is based on considerations that physicists thought conclusive. It is extremely unlikely that they should go back on this[4]; too many brilliant successes have already been scored by the new quantum theory, which is the synthesis of the corpuscular and undular theories. It is rather to be expected that further drastic modifications in the prevalent scientific outlook will prove to be necessary in the near future (cf. Jordan, 1959).

It will now be evident that the foundational crisis in the exact sciences – the example discussed above is only one of many cases – in no sense justifies misgivings as to the future of science. On the contrary: abandonment of principles once considered unassailable is the price to be paid for the progress of science.

This view is in turn supported by the outcome of historical considerations. The principles which must now be gradually abandoned practically without exception find their origin in the oldest period in the history of science. It is quite understandable that man's first scientific insights, which formed such a striking contrast with the mythical notions prevalent till then, were enthusiastically credited with absolute certainty. As these insights were confirmed by the outcome of increasingly advanced research, the conviction of their being unassailable inevitably grew stronger, even though attempts to account for this unassailability by a theory of knowledge failed to have much success. Thus the 'foundational crisis' is really a crisis of prejudices, albeit respectable ones. In the 16th to 18th centuries and, to a somewhat lesser degree, in the 19th century, the exact sciences were steeped in Greek – sometimes even much older – conceptions, which are now being gradually abandoned. Now for the first time a western science is developing. In other fields of civilization an analogous development can be discerned.

I shall now pass on to the third part of my argument and discuss *contemporary phenomena in politics and society*. I repeat that I want to consider these phenomena from the stand-point of exact science. I also wish to point out that some restraint is required here which was previously not necessary.

The phenomena to be discussed can be classified into three categories: *events, institutions* and *ideas*.

In this essay I wish to deal principally with social and political ideas, to consider institutions only in so far as this is required by our discussion of the ideas, and to leave events altogether aside.

Anyone sufficiently familiar with the exact sciences is immediately struck by the fact that the development of social and political ideas is much slower, and that it is much more strongly influenced by (group) traditions and to a much lesser degree by experience. I propose to attempt an explanation of this phenomenon first, and then to point out the dangers attending the slow development of social and political ideas.

In order to explain this slow development of social and political ideas I should like to start with the following observations:

(i) Novel ideas in the exact sciences are taken seriously and may have consequences as soon as they have gained a hearing with a small group of leading specialists. This is not, as a rule, the case with new social and political ideas. These must be popular amongst a very wide public, which requires prolonged and intensive promoting, probably involving serious dangers for the advocates.

(ii) Social and political ideas practically always concern existing institutions – no matter whether these are described objectively, attacked or defended. Institutions, however, usually change extremely slowly.

Point (i) speaks for itself; as for the slow development of social and political institutions referred to under (ii), I should like to offer the following explanation:

(a) Rapid changes in institutions are in general not expedient, since they detract from the respectability and thus from the authority of these institutions. The slow development of marital law for instance is evidence of the respectability of matrimony. This may help to explain why the practical effect of revolutions and unduly radical reforms is usually disappointing. The abolition of slavery is a case in point: much as moral considerations demanded it, still the social position of those who were to profit by it seems at first to have been little improved.

(b) Most institutions tend to favour one group above another (if this is not the case, there is no incentive whatsoever to change matters, if we leave aside the possibility of a *coup d'état* by criminals without political object). The second group will be inclined to change the institutions, while the first will try to defeat this purpose. The privileged position of the first group will of course lessen the chances of the second to realize its aspirations, as is again illustrated by the long continuance of slavery.

One of the ways in which a ruling group may try to maintain its privileged position is by influencing social and political ideas in a conservative sense. To achieve this, the ruling group will also try to influence the work of scientists.

7

In the social and political sciences the scholar's inevitable social engagement has a far more harmful influence than it has in the exact sciences. The manager who commissions a chemist with research work expects a scientifically objective result, which he may eventually use in his factories. If so, he will prevent the discovery becoming known prematurely, but he is not interested in spreading chemical falsehoods. The politician, however, expects of the economist he is protecting to defend the policy he has pursued or wishes to pursue. He will oppose the divulgence of ideas that may be detrimental to his propaganda, even in specialized journals.

(c) Even progressive groups, usually formed by those who consider themselves slighted or injured by existing institutions, will impede the development of social and political ideas. For even within a progressive group there are leaders who will want to perpetuate their relatively privileged position by the means discussed under (b). Besides, the very fact that a group is in a disadvantageous position means that it will at least try to preserve its ideological unity at all costs. This is a check on the development of the ideas which is much less noticeable within established groups.

Obviously this is a vicious circle. The slow development of social and political institutions impedes the development of ideas; the slow development of ideas hampers a rapid development of institutions. The consequence may be that our ideas and our institutions change more slowly than would be desirable in view of the needs of a society that is to function adequately.

This whole argument leads us, as far as the third part of our inquiry is concerned, to a pessimistic conclusion. It is to be feared that the rapid development of the exact sciences, and consequently also of technology, together with the slow development of social and political ideas and institutions, will lead to an ever-increasing disparity between the two. If we contrast the atomic bomb with the Security Council, it becomes evident that this lag could mean disaster.

Yet even here we should be warned against an unrestrained and paralysing pessimism. Indeed the threat of a catastrophe sparing neither privileged nor underpriviliged groups augments the chances of breaking the vicious circle enclosing our social and political ideas and institutions before it is too late.

It will now be clear where I would look for the solution to the most oppressive problems first. A scientific institute should be founded that is in a position to pursue social and political investigations independent of political bodies and of ideologies and to make the results of its investigations known without any restrictions. The members of this institute would have to be guaranteed financial independence and personal immunity. The international status as enjoyed by the Vatican would perhaps be the most appropriate one for this institute.

Having arrived at the end of my reflections, I should like to formulate my conclusion as follows. Cultural fatalism has proved to be ill-founded; a foundational crisis does in fact exist, but this does not justify pessimism; though it must be admitted that our world is confronted by social and political phenomena which are to be considered deeply alarming. Still, I wish to caution once more against unrestrained pessimism.[5]

Furthermore, I wish to voice my conviction that it is not in the first place necessary, or even desirable, to find out who is guilty of the social and political evils of our time. As Spinoza would have said: I wish these evils were studied just as straight lines and circles are studied in geometry, instead of being deplored or made fun of.

Social and political life, it seems, is in the last resort governed by common and conflicting interests. The distinction between self-interest and the interest of others, so prominent in ethics, is of minor importance in this connection, since the social effect of the promotion of interests is not affected by it – the case of a trustee in charge of the property of a charitable institution illustrates this quite well. If in spite of all conflicts a certain social order is maintained, this proves that community of interests prevails in the end, in spite of all self-interest. This may lead us to face the future of our civilization with some confidence; it does not, however, relieve us of the obligation to make every effort for the survival and further development of our civilization. This effort, if it is to be successful, must be guided by rational deliberation, provided it is based on experience.

REFERENCES

[1] Spengler found a supporter in the mathematician C. J. Keyser (Keyser, 1947).
[2] Francesco Maurolico, *Arithmeticorum libri duo*, Venetiis 1575. The importance of Maurolico is rated less highly by Freudenthal (1953).

9

[3] Blaise Pascal, *Traité du triangle arithmétique*, Paris 1665.
[4] A few years ago, Bohm made an attempt to eliminate the principle, but his proposal has met with little approval (Bohm, 1952a, b, 1953).
[5] In 1947 I read a newspaper article by Geyl confirming me in this attitude. Cf. Gey (1958).

II

NATURAL SCIENCE, PHILOSOPHY, AND PERSUASION

Before entering upon the proposed topic I have to make a few terminological remarks.

(1) The term 'natural science' is used in a broad sense; it is taken as including technology. Indeed much that is said about natural science applies to many other domains of science as well: not only to pure and applied mathematics but also, e.g., to economics, which is becoming more and more like an exact science.

(2) The term 'philosophy', however, is to be taken in a somewhat restricted sense. I have only applied it to those, in a broader sense philosophical, discussions which, according to certain standards, are sufficiently systematic, scientific and logically consistent.

(3) The term 'persuasion' probably requires most explanation. Whereas natural science and philosophy are public in character, a persuasion is entirely personal in two respects. Firstly, we may, of we should wish to, refrain completely from studying science and philosophy, but we would still make use now and then of the achievements of others in these fields. Everybody, however, whether he likes it or not, *has* a certain persuasion, so that it is hardly possible to fall back on someone else's persuasion. Secondly, science and philosophy are as it were independent public institutions which are available whenever we should want to use them. A private persuasion, however, is part of the personality and thus comparable with a person's hand-writing, his countenance or his character; we cannot disown it without belying our own nature.

We could perhaps define a persuasion somewhat more precisely as a set of convictions, valuations, attitudes and aspirations, determined by the intellect as well as by the emotions and the will, and which can to a certain extent be expressed in language intelligible to all.

A person's persuasion is formed in the course of his own life, partly by environment (upbringing, religion, school, politics), partly by personal experiences (contact with the opposite sex, professional activities) and, provided that a sufficient degree of intellectual maturity has been attained, partly by one's own conscious effort. Therefore no two persons have exactly identical persuasions.

However, a large number of people quite often have *more or less* the same persuasion, at least in so far as it is expressed intelligibly – though this naturally applies mainly to those who are less educated or less experienced. These widely shared persuasions often strike one as rather stereotyped. In this sense, then, we can say there are certain types of persuasions and even groups sharing a certain persuasion. Yet extreme caution is called for here, since especially in the case of less educated or less experienced people essential parts of their persuasion are often not expressed in words or even fully realized. We might in these cases compare a persuasion with an iceberg, only a small part of which is visible, while the greater part remains hidden under the surface of consciousness. Occasionally, to be sure, these hidden convictions manifest themselves in an intelligible way, namely in the behaviour of the person concerned. There are cases when somebody emigrates or becomes a conscientious objector, although outward circumstances do not seem to supply a clear motivation.

Religion is undoubtedly one of the environmental influences which, besides upbringing, strongly influences a person's persuasion. This does not mean, however, that this persuasion is *determined* by religion; for the eventual effect of the influence exerted by religion essentially depends on the way in which certain religious convictions are assimilated by the particular person concerned. Certainly, it is to be expected that a certain persuasion will be strong and even prevailing among those professing a certain religion. Still, even convinced believers sometimes have strongly dissenting persuasions; this will be discussed later on. As a rule, of course, if somebody breaks with a certain religion this *is* the result of a conflict between received religious doctrines and personal convictions.

A further determinant in the development of a person's persuasion is the study of science; in particular we are concerned with the effects of studying natural science. As in the case of religion, the study of science is conducive to the development of certain types of persuasion, though this does not imply that it *determines* the scientist's convictions. This becomes clear when we realize that among the most prominent students of natural science, and even within a special field of science, there are men with strongly divergent persuasions.

12

This is the opportunity to reflect briefly upon the alleged conflict between beliefs and science, an opportunity which I do not want to miss. We must make a distinction between several radically different situations.

First there is the possibility of a conflict within the person. A fine example is the case of Georg Cantor (1845–1918), who constructed his theory of sets with the explicit aim of giving a mathematical theory of the actual infinite. Thus he rejected the traditional view according to which only the potential infinite could be dealt with by mathematics. This is a point which requires some elucidation.

Given a circle, we may inscribe and circumscribe it with regular polygons of, say, 4, 8, 16, 32, ... sides. Once the number of sides has become very large, the circle cannot practically be distinguished any more from the inscribed and circumscribed polygons. One could say now that a circle is a regular polygon with infinitely many sides, and an attempt might be made to calculate the area and circumference of the circle on the basis of this principle. As we know, however, this is not a sound method.

The correct method is to show that, if the number of sides increases indefinitely, the circumferences of the circumscribed and inscribed polygons *approach a common limit*, which apparently can be considered the circumference of the given circle. The same holds, *mutatis mutandis*, for the respective surfaces.

The first case was a recourse to the actual infinite, while in the second case we content ourselves with a recourse to the potential infinite. For in the second case we do not introduce a polygon with infinitely many sides; we only find out what happens to a polygon if the (finite) number of its sides is allowed to increase indefinitely.

Even in Antiquity it was concluded that since a correct treatment is impossible in the first case, the actual infinite is altogether impervious to mathematical treatment. All kinds of philosophical and theological speculations were based on this conclusion. The view was finally discarded by Cantor.

It appears from a number of facts that there was a conflict on this issue between Cantor's personal persuasion and his scientific endeavour. I only wish to mention that Cantor dealt minutely with past philosophical arguments in his writings and that he even saw fit to seek the advice of theologians.

13

This case seems to me particularly instructive for the following reasons.

(1) The conflict is real. Since the problem originated in mathematics, the relevance of recent developments in this discipline is not a matter of doubt.

(2) There is no power conflict, as in the case of Galilei or of Darwin, so that the problem can be discussed dispassionately.

(3) The problem is wholly theoretical, so the matter cannot be decided by experience.

The conflict between science and persuasion plays a much less prominent part in the polemic between Couturat, Poincaré and Russell round about 1900–1910. From this battle mathematical intuitionism has emerged, but the great majority of mathematicians has accepted Cantor's constructions and makes extensive use of them. Some believing thinkers however, Catholics and Calvinists alike, continue to raise objections to Cantor's ideas.

In the case we have just considered there was a conflict within the person. A conflict between science and creed or persuasion may however manifest itself in a very different manner. In this connection it is interesting to compare the situation that has resulted from Cantor's ideas with, on the one hand, the reactions following discoveries made by Galilei, Newton, and Darwin, and, on the other hand, those produced by publications of Einstein, Bohr and Heisenberg. The discoveries made by Galilei, Newton, and Darwin influenced the way of thinking of whole generations so profoundly that eventually a power conflict arose; the publications by Einstein, Bohr, and Heisenberg, though they did give rise to a polemic that was often conducted very fiercely, did not result in a power conflict and there was no question of any serious personal conflict of conscience. It would seem that conflicts between beliefs (or persuasions) and science are becoming less and less acute.

Strongly increased specialization is no doubt an important factor in this process. The philosophical and religious problems raised by Darwin's work were such that any more or less educated person could understand the essentials. The problem raised by Cantor's ideas is much more abstract, so that only very few people are interested in it. Even these few, however, nowadays do not have the time or the energy required for profound reflection upon philosophical problems. As a result, intellectual life is much less exciting in our time than it was, say, a century ago.

The reader might now try to entangle me in my own argument by raising the question: what difference can it make how much time we spend in forming or reflecting upon our persuasion, if everybody, whether he likes it or not, already possesses a certain persuasion to begin with? This is not, however, a sound objection. We cannot say: it does not matter *what* persuasion we have, provided only we have one. Indeed, *what* we do, *how* we react when we are confronted with perhaps disastrous developments and consequential decisions depends to a great extent on our persuasion. If our convictions should be insufficiently developed, we may fail altogether to arrive at a decision – a shortcoming which is not at all infrequent among contemporary men of science. If our persuasion should lack proper foundations, we may reach a decision, but it might be one which, rightly or wrongly, we shall come to regret later on.

As I observed above, our persuasion, though influenced by numerous external factors, is not determined by them. At this point I must add that our persuasion cannot be and should not be dictated to us by others. The reason is that the problems that may arise in our lives and the decisions we must make are shared only to a small extent with other people and are otherwise highly personal. A persuasion which has been dictated to us by others may fail us in the very situation where we need it most. Furthermore, we ought not to prescribe anybody a persuasion, because this burdens us with a responsibility that we cannot take upon ourselves.

Thus the predicament of the modern scientist is almost desperate. He is not given the opportunity to construct a reasonable system of beliefs that fits his personality. Nor can it be dictated to him by others; this is undesirable not only by virtue of the general considerations mentioned above, but also because the demands made upon the persuasion of the scientist are exceptionally high. And finally, the modern scientist is more likely than his predecessors to be confronted with situations that he will be able to manage only if he has a well founded system of beliefs.

In a sense, philosophy will be able to help him out of this predicament. In saying this, I am faced once again with the possibility that the tables will be turned on me: why should philosophy be granted that which has just been denied other spiritual powers?

As before, the argument fails. Philosophy cannot and should not aim, at constructing a ready-made persuasion for everybody. It does however, though on an abstract level and *sub specie aeternitatis*, deal with all the problems that also arise in our private lives if we try to develop a well founded system of beliefs. In so far as this is possible, philosophy also provides objectively tenable solutions for these problems; otherwise, we would not be justified in speaking of *scientific* philosophy.

A much more important facet is, however, that philosophy supplies a more or less complete survey of all those problems with which we may be confronted in our private lives at some time or other, and of the various view-points we can take with regard to these problems. By considering these problems and view-points with the help of this kind of survey even before life forces them upon us, we can, at least to a certain extent, prepare ourselves for decisions that we might be called on to make in the future. Although the study of philosophy may lead to conflicts, this only applies in so far as it makes us aware of lacunae in our persuasion.

I now want to discuss some of the problems which might become acute for the natural scientist of today.

(1) We should continue to be prepared for developments within science which will fundamentally change our picture of the world and possibly our conceptions of man. Any such developments will involve far-reaching consequences, so much so that, notwithstanding the opposite tendency mentioned above, a power conflict will arise, forcing the scientist to take sides. In order to be prepared for such developments, we should certainly not, as is frequently recommended, make light of the ideological consequences of developments in the past.

I repeat that the problems with which we are concerned here are also considered in philosophy, though on a general level, not as private problems. I think it is useful to repeat this, since on several occasions it has been claimed that these problems belong exclusively to theology. Some years ago there was a discussion among Dutch philosophers and theologians on whether philosophy or theology could claim a central place in the universities. According to an opinion voiced by a prominent theologian, *the eschata* (defined as 'the ultimate data and structures') *of all sciences are of a theological nature.*

It is no doubt quite justified to say that theology, like philosophy, touches upon the very principles of the various sciences. Apparently, however, it is also claimed that the inquiry into these principles must be viewed as a specifically theological affair. For reasons of principle as well as for practical reasons I think there are serious objections to this.

(i) Almost everywhere in the world the study of theology is, even at the universities, the specific affair of the churches. The study of the deepest principles of the various sciences, each claiming universal validity in a particular field, would thus be kept strictly within the sphere of the churches of various denominations and would be firmly linked with certain religious beliefs. It is obvious that this is not advantageous for the development of this study; it is very probable that it would result in numerous conflicts, as we may learn from the history of science.

(ii) The theological view quoted above is based on the assumption that the study of the principles of a scientific discipline can be pursued independently of the actual development of the discipline. For example, natural science uses concepts such as space, time and causality. These concepts are firmly established, so that a physicist may make use of them without bothering about their justification. On the other hand, this implies that his research can never lead to results that would affect the universal validity or applicability of these concepts. Consequently, the study of the concepts mentioned cannot be pursued along specifically scientific lines. This study thus belongs to a quite different branch of knowledge. Aristotle, who was the first to defend the view on the study of principles which I have indicated summarily here, allocated this study to metaphysics, which, although it remained formally independent, later became more and more a part of the study of theology. Aristotle's viewpoint has however been superseded by the development particularly of the exact sciences during the last 100 or 150 years. We have come to see that a science may reach a stage where a drastic revision of even its most fundamental principles becomes necessary. This first became apparent in mathematics, as it did in physics later on, and ever since the realm of principles has been a place of unrest. It will be obvious that the problems arising in this connection are almost completely unrelated to theology and that they must be considered in close connection with the research taking place in the discipline concerned. The systematic discussion of problems such as these is the principal task of the philosophy and the methodology of the exact sciences.

All these problems, however, are theoretical, whereas in recent times natural scientists are concerned even more about problems which must be considered ethical. I wish to discuss two such problems.

(2) May we take part in scientific research that will, if applied in practice, probably provide the means to destroy all human life? May we take part in the development of this kind of applications? Can we refuse to co-operate, even if the interests of our country or the survival of civilization are at stake? Should we make our co-operation depend on whether we have a voice in the decision on any possible use of such exterminatory devices? Are we in a position to bear the responsibility involved in our acceptance to have a say in these matters?

I could not and should not suggest a ready-made answer to these questions. I am however justified, indeed this is a philosopher's task, in pointing out the possibility that at some time in his life a scientist will be seriously confronted with these problems. He will than have to be able to make responsible decisions without delay, decisions which will be possible only on the basis of a well founded system of beliefs. In the absence of this there will be a lack of determination which in recent decades has become the tragedy of many a scholarly existence.

(3) Perhaps the prospect of a society directed wholly technocratically is even more oppressive. Scientific technology will comparatively soon supply the technical *means* for this *dirigisme*, but it is no part of its task to furnish the objectives as well. Nor, extraordinarily, can the social sciences provide reasonable and generally accepted objectives. So the question arises whether under these circumstances the natural scientist is justified in contributing to the development of the technical devices in question. The risk of a wholesale annihilation of human lives may be less grave in this case, but there is a danger that human life will be robbed of its actual 'significance' on a very large scale. It would be wrong, of course, to attach too much value to negative utopian prophecies like Huxley's *Brave New World* and Orwell's *1984*. Nevertheless we shall be well-advised not to have too cheerful an opinion of a wholesale technocratic *dirigisme*.

I will now sum up and state my conclusion. Every future natural scientist ought in the first place to see to it that he masters the subject of his choice thoroughly. Specialized studies require the constant effort and dedication of university students, and the knowledge acquired at the university will have to be kept up to date, supplemented and assimilated more profoundly after graduation. It is however not to the advantage of a natural scientist if he only acquires an extensive and profound knowledge of his own subject and, maybe, of some adjoining subjects. Nor is society so easily satisfied, even if we restrict our attention to the comparatively narrow confines of our normal professional activities. Are we not all in due time confronted with problems which nobody who possesses only a mature character and a thorough training for his profession can ever hope to cope with? In such situations there must be a well founded and justifiable system of beliefs to rely on. The foundations of such a system of beliefs are laid during one's up-bringing, while school, church and professional life may each contribute in important ways.

However, on the level of responsible scientific research, a persuasion acquired while the person concerned remained passive does not usually suffice. The prospective scholar will have to make an active contribution by doing his own thinking about the problems which could arise in the course of a human life, whether in connection with his profession or otherwise. In particular he will have to give thought to those questions which might become pertinent especially to him. As we know, excellent opportunities are offered for this during college life.

In the usual university courses, natural science students are not confronted with the kind of problem which we have considered here. The case is different for students of law or divinity; however, in view of the structure of modern society and of the place which natural science occupies in it, natural scientists are no less concerned about these problems than lawyers or theologians.

The study of philosophy, even on a moderate scale and on an elementary level, could assist the prospective natural scientist in developing his own persuasion. Philosophy cannot furnish ready-made answers for immediate use for all possible vital questions. On the contrary, philosophy rather removes the most urgent questions from the private sphere and raises them to the level of generality, so that it can, while taking philosophical tradition into account as well, view the problems under different aspects and point out connections between the various questions. Indeed we may already feel relieved to learn that the very problems that we are struggling with ourselves, or at least closely related ones, have already been reflected upon by the greatest minds of the past.

I certainly do not want to claim a monopoly for philosophy, nor do I have any intention of making a plea for the introduction of a compulsory general elementary course in philosophy; in my opinion the value of such an elementary course is rather dubious. Only those who feel a calling towards philosophy and are capable of thinking independently will benefit from the study of it, and no others should be advised to choose it as their subject. Still, we should not wait until we are so deeply involved in the problems – and perhaps even in internal conflicts – that we can find no way out. Then, of course, we shall no longer be able to keep aloof, as is required in an independent philosophical consideration of the problems. For as I have argued, philosophy does not supply ready-made answers for immediate use; its help would thus be too late. There is even a risk that starting with philosophy at the wrong moment will aggravate existing inward conflicts.

In conclusion I wish to consider the following question, closely connected with the above. Some people appear to be under the apprehension that the study of philosophy could lead to 'unscientific' thinking habits. In my opinion this danger only exists for someone who has a certain degree of intellectual instability even before he embarked upon the study of philosophy. The intention and the essence of philosophy are genuinely scientific. It is not just the expression of the philosopher's personal persuasion; like all other sciences, philosophy is concerned with generality. Its statements and doctrines are claimed to be recognized as true, universally valid and objectively founded. For this very reason philosophy must expressly relinquish every pretension of prescribing anybody a certain persuasion, whereas it does claim to be able to provide anybody who turns to philosophy with the materials which he must compose into his personal persuasion by his own independent reflection. How can these two contrary statements be made to agree? This is a philosophical problem which I should like the reader to reflect upon for himself.

III

METAPHYSICS AND SCIENCE

The traditional metaphysical and epistemological systems were evolved in an attempt to solve certain problems that are meaningful within Plato's and Aristotle's conception of science. As a result of the modern development of science, this conception, which was taken for granted for many centuries, is no longer tenable. It has become clear that particularly the evidence postulate (cf. p. 76) – i.e., the condition that the fundamental concepts of a scientific discipline must be clear without need of explanation, and that the fundamental principles must be acceptable without need of proof – is no longer accepted by modern science. Consequently, a new conception of science must be envisaged, while traditional metaphysics and epistemology will have to make way for modern style foundational research.

This argument is clearly based on the assumption that the metaphysics and epistemology of antiquity were felt to be scientific in character. In support of this assumption, I wish to pause for a moment to discuss the views of some of the greatest classical philosophers concerning the relation between wisdom and science.

In the *Theaetetus* (145e) Plato identifies wisdom with science, while in the *Republic* we find him considering the study of science (in particular mathematics and astronomy) as an indispensable training for the study of philosophy.

The scientific background of Plato's own doctrine is the logical, mathematical and astronomical research which he did at the Academy, together with Aristotle, Eudoxus, Theaetetus, and others. Although the results have not come down to us in their original form, it is possible to reconstruct them fairly reliably out of the writings of Aristotle, Euclid, Sextus Empiricus, Proclus, and others, with the aid of some hints in the dialogues. This reconstruction gives us a clear idea of the main points of Platonic philosophy and puts its scientific nature beyond doubt.

Yet even in late Antiquity the scientific nature of Plato's doctrine was no longer understood (Milhaud, 1900, pp. 195–200), nor does it, as a rule, receive sufficient attention in our days.

This lack of comprehension of Plato's scientific intentions is partly due to the form in which his ideas have been handed down to us. The dialogues are 'exoteric' writings: they were not meant for those who had received a scientific training at the Academy, but rather for the large number of laymen interested in philosophy. Thus a clear understanding of Plato's ideas cannot be gained from studying only the dialogues, just as one cannot expect to become familiar with, e.g., Eddington's ideas by reading only *Stars and Atoms* and *The Nature of the Physical World*.

The exoteric nature of the dialogues is manifest in the dramatic framework, in the insertion of 'myths', in the often downright quibbling arguments that Socrates puts forward against his antagonists in the course of the discussions, and also in the fact that the problems raised remain unanswered so frequently.

Let us by all means not suppose that Plato raised problems in his dialogues which he could not solve to his own satisfaction. If he so often fails to provide a solution, this is rather due to the consideration that this solution presupposed a dialectical training which he could not expect his readers to possess.

We should not, therefore, take the dialogues too literally. On the other hand, this reservation should not lead us to deny Plato firmness of convictions. In the *Republic* (534b-c) Plato clearly and with unmistakable depreciation states his opinion of people who lack firm convictions:

Until the person is able to abstract and define rationally the idea of good, and unless he can run the gauntlet of all objections, and is ready to disprove them, not by appeals to opinion, but to absolute truth, never faltering at any step of the argument – unless he can do all this, you would say that he knows neither the idea of good nor any other good; he apprehends only a shadow, if anything at all, which is given by opinion and not by science; – dreaming and slumbering in this life, before he is well awake here, he arrives at the world below, and has his final quietus. (Jowett's translation.)

Indeed, Plato himself makes it clear that the more subtle elements, the real core of Platonism, cannot be adequately discussed in the dialogues (see *Phaedrus* 275c ff.). He goes on to argue that only an oral exposition could serve the purpose; and he actually gave such an exposition at the Academy. Important fragments of his discourse *On the Good* have, more or less by chance, come down to us – though little attention has been paid to them until recent years – and it is these fragments, together with other data, which enable us to form an idea of Plato's thought which is much more correct than the idea conveyed by the dialogues alone.

In consequence of this, some views on Plato that have become more or less traditional have to be drastically revised. Whereas Plato used to be considered as an *esprit fin* in the sense of Pascal's famous distinction, it now appears that it would be more correct to stamp him as a *géomètre*. Plato aims at a level of abstract reasoning which could only be said to have been really attained quite recently. Thus his ideas on the foundations of mathematics form the starting-point of a historical development which leads from Apollonius, Leibniz and Bolzano to Frege, Cantor and Russell. It is noteworthy that Leibniz, Bolzano and Russell share Plato's interest in social and political problems.

Though more recent investigations have led to results deviating, even in essentials, from Plato's views, and though this means that these views might easily strike the modern expert as somewhat primitive, this should not, of course, detract from our admiration for Plato's genius. But neither ought this circumstance to be used as an argument against the view that these modern thinkers are the legitimate heirs of the tradition originating with Plato.

For Aristotle too metaphysics is scientific in character. He defines mathematics, natural science and theology (i.e., metaphysics) as "theoretical philosophies" (*Metaphysics* E i, 1025ᵃ10 ff.), and wisdom as "the science concerned with first causes and principles" (*Metaphysics* A i, 981ᵇ25 ff.).

A similar view was expressed by Descartes many centuries later (*Les principes de la philosophie*, préface):

The whole of philosophy is like a tree: metaphysics is the roots, physics is the trunk, and the other sciences are the branches.

This will have to suffice here as a demonstration of the scientific character of metaphysics.

In addition to this purely scientific tenor, however, there were also other, religious traits in the old metaphysics. It should be noted that it is only possible to distinguish these two trends in metaphysics from a modern view-point. The distinction will hardly have existed for the ancient mind.

The religious significance of the knowledge of first principles, as

23

provided by metaphysics, may be discerned in two well-known Latin verses:

Felix qui potuit rerum cognoscere causas (Vergil).
Nihil mirari prope res una quae possit facere et servare beatum
(Horace).

Knowledge of that which underlies everything gives true happiness, unshakable peace of mind, by eliminating the wonder at our personal fate.

Also in Plato and Aristotle we find the view that wonder is the starting-point of all philosophy. It is Plato, however, who reveals the background of this seemingly rather trivial and not very profound idea. For according to him philosophy is the "practice of death" (*Phaedo* 81a).

If we want to ascertain the meaning of these curious pronouncements, I believe we shall have to relate them to certain views that were widespread in Antiquity. The Ancients knew mysteries in which the mystic underwent death and was subsequently raised to eternal life. According to the ancient views, immortality was acquired by this initiation; death could no longer touch the initiated.

Knowledge of the first principles, then, should apparently be considered an initiation, for which one is prepared by wonder and which is therefore to be looked upon as an intellectual death. Knowledge of the first principles renders one immune from wonder, from intellectual death, and in this way intellectual immortality is acquired.

Thus a philosopher, a metaphysician, is an initiate, a soul reborn into intellectual immortality, who by his insight into the deepest nature of things has triumphed over intellectual death, over wonder.

At the same time this view is the key to the explanation of several peculiar pronouncements of some ancient authors which are otherwise difficult to understand. For example, Aristotle's doctrine according to which the νοῦς is immortal; it is the νοῦς which supplies man with knowledge of the first principles and thus with intellectual immortality. Or Chrysippus' doctrine, which attributes an immortal soul only to the wise, that is, to those who are intellectually reborn.

In modern times the idea of an intellectual rebirth is again to be found in Descartes' words (*Les principes de la philosophie*, i, 1):

we should, once in our lives, call everything into question;

or, even clearer perhaps, in Spinoza (*Ethics* v, Proposition xlii, Scholium):

The sage is hardly moved in his soul, but in the awareness of an eternal necessity of himself, of God and of things never ceases to be but always possesses true peace of mind.

Only when related to the idea of intellectual death and rebirth does the import of the evidence postulate become quite clear. If knowledge of first principles is to furnish an effective remedy for the wonder at our personal fate, which is experienced by anyone who is in the habit of thinking, then this knowledge must be justified to his reason as perfectly final, not subject to any revisions or corrections.

The religious import of metaphysics explains why, even in our days, so many people appear to be attracted by it. Even today, the true metaphysician imagines himself to be an initiate, existing at an exalted level high above the rest of humanity. It is therefore quite understandable that the development of science, in which the evidence postulate and thus also the scientific value of metaphysical reflection became dubious, should have given rise to violent reactions.

The tactics used in defence of the scientific import of metaphysics and of the accompanying epistemology consist of two manoeuvres. In the first place war is declared directly upon all modern scientific theories that are incompatible with the evidence postulate. As an example, I mention the criticism, often of an extremely violent kind, that has been levelled against symbolic logic, non-Euclidean geometry, relativity theory and the newer quantum theory. I do not wish to discuss this criticism here; in my opinion it has only resulted in discrediting even more not only metaphysics, but philosophy as a whole, especially in scientific circles.

In the second place it is held that in addition to modern science – often qualified somewhat disdainfully as 'specialized' in this connection – there are disciplines of a different kind, referred to as 'philosophical disciplines', which are to serve as the starting-point for metaphysics (cf. p. 30). These 'philosophical disciplines' are a suitable starting-point for metaphysical reflection since they agree with Plato's and Aristotle's conception of science and in particular with the evidence postulate. Thus, it seems, the scientific import of metaphysics may be salvaged.

However, as soon as we inquire what exactly is offered us under the impressive title of 'philosophical discipline', it at once becomes clear that this is an illusion. We find that the title more or less covers a collection of ideas that have been abandoned by 'specialized science'. Thus Aristotle's teleological physics, Descartes' corpuscular physics, Newton's physical doctrine of action at a distance live on to this day as 'philosophy of nature'.

This implies, of course, that these 'philosophical disciplines' and the metaphysics based on them do not supply explanations of every-day experience. The scientific theories of Aristotle, Descartes and Newton, for example, have been abandoned by 'specialized science' for no other reason than their being incompatible with the outcome of scientific research.

This line of reasoning is usually countered with an argument of the following kind. Philosophy is independent of physical explanations of physical particularities, since it is based on the great and general facts that may be found everywhere. By its very universality, the object of philosophy cannot be dependent on special theories of natural science.

This counter-argument undoubtedly enjoys a certain popularity, and it certainly cannot be denied a semblance of soundness. For example, relativity theory, so glaringly in conflict with judgments that are thought to be self-evident, such as the notion of simultaneity, is justified, as against classical physics, by experimental results which are qualitatively not very striking and are very small in number. So when we are constructing a metaphysical system – since its object is to eliminate the wonder at our personal fate, why should not we simply ignore these experimental results and the accompanying theory and accept only the evidence of absolute simultaneity?

In reply to this counter-argument, both scientific and more general considerations are relevant. In science it is not warranted to take only 'great and general' facts into account and to discard experimental results like those just referred to as 'unimportant'. For the scientific significance of a fact depends not only on its prominence or on the frequency of its occurrence but also, in certain circumstances, on the extent to which it determines the course of scientific thought.

The more general consideration is that by discarding less conspicuous experimental results, we would be wilfully reducing the domain of our experience, thus deliberately confining our outlook to the sphere of everyday experience. Attractive as such a deliberate restriction of our intellectual outlook may be for some, it certainly is not in the spirit of Aristotle, the founder and great master of traditional metaphysics, who regarded progress in philosophy as a gradual extension of this outlook:

Philosophy arose then, as it arises still, from wonder. At first men wondered about the more obvious problems that demanded explanation; gradually their inquiries spread farther afield, and they asked questions upon such larger topics as changes in the sun and moon and stars, and the origin of the world. (*Metaphysics* A ii, 982 b 12, John Warrington's translation.)

Indeed an appeal to 'philosophical disciplines' can only lead to pseudo-scientific, never to scientific explanations of experience.

Thus our conclusion must be that a metaphysics which traditionally combines a religious and a scientific import, on the one hand satisfying the evidence postulate, thereby eliminating once and for all the wonder at our personal fate, initiating us in the mysteries of life – solving the 'problem of life', as the phrase goes –, and helping us to triumph over intellectual death by obtaining intellectual immortality, while on the other hand fulfilling the usual scientific requirements, such a metaphysics must be considered impossible in view of modern scientific development.

What then should be the task of scientific philosophy? In the preceding essays I have already mentioned the investigation into the principles or foundations of the various sciences. In mathematics, for example, this research has already taken impressive proportions.

In addition to this, however, there is room for a scientific philosophy of life, seeking to explain human experience on the basis of the outcome of scientific research.

It could be objected that the task of this philosophy of life would coincide with the task of science as a whole. However, it seems reasonable to expect that, e.g., psychoanalysis can do more towards explaining our behaviour than, for instance, astrophysics. This does not mean that a scientific philosophy of life may discard astrophysical results. It does mean, however, that the results of the various sciences can be grouped together in such a way that the main attention is focused upon that which is important for us *qua* human beings.

This philosophy of life would have to be considered applied rather than pure science, as its objective lies outside science as such. This objective is to satisfy man's need for an explanation of his experiences. Nevertheless, like any applied science, a scientific philosophy of life could influence and stimulate scientific research. By grouping together the results of this research from its own philosophical view-point it could contribute towards the discovery of territory that has not, or not sufficiently, been explored by science. Under certain circumstances its influence could become harmful, since man cannot remain disinterested when his own personal fate is at stake.

The essential difference between a modern scientific philosophy of life and traditional metaphysics is that modern scientific philosophy would never fail to realize that its explanation of human experience will always and inevitably be incomplete and tentative.

IV

SCIENTIFIC PHILOSOPHY: ITS AIMS AND MEANS

In contemporary philosophy, divergent though its manifestations may seem, three main currents can be clearly distinguished: traditional philosophy, irrationalism, and scientific philosophy.

By *traditional philosophy* I mean any philosophical doctrine which derives its main tenets from a philosophical system established in the past; consequently, Platonism and Aristotelianism as well as Kantianism, Hegelianism and Marxism fall under this notion. At first sight, it may seem arbitrary to subsume such divergent doctrines under one heading; however, it may be shown that in spite of their divergence they have quite a number of typical features in common. For instance, they all start from Aristotle's conception of science.

The term *irrationalism* needs no further explanation, as I take the word in its generally accepted sense. During the first decades of this century it was typically represented by Bergson's vitalism, by James's pragmatism, and by German philosophy of life as represented by, e.g., Ludwig Klages. At present, the different forms of existentialism are the most important manifestation of the irrationalist mentality.

Scientific philosophy is characterized by its ambition to deal with philosophical problems in a strictly scientific manner. By philosophical problems I mean the sort of problems usually discussed by philosophers. However, a distinction should be made. There are a number of problems – regarding the principles of knowledge, the standards of valuation, the nature of man – which present themselves unavoidably to any philosopher; besides these primary problems there are secondary problems which depend on the special method adopted in dealing with the primary ones. These secondary problems are of a more technical nature; they may be important for one school of thought and devoid of interest or even non-existent for other schools.

Though scientific philosophy attaches no great value to tradition, it may in some respects claim Democritus, Plato, Aristotle, Descartes, Locke, and Leibniz as its forerunners. The sharp distinction between traditional and scientific philosophy is of a very recent date. The distinction originated with the modern development of science, which led to results incompatible with fundamental assumptions of traditional philosophy.

For example, the so-called *evidence postulate* is such an assumption (cp. Scholz, 1930/31, and p.76). According to traditional philosophy, science has to start from principles which are self-evident and not subject to revision. Logical laws and geometrical axioms are the customary examples of such principles. Traditional epistemology, as established by Plato, Aristotle, Descartes, and Kant, aims at explaining our knowledge of these principles. Modern science, however, no longer accepts the evidence postulate; this appears from the development of many-valued logic, of non-Euclidean geometry, of the theory of relativity, of quantum mechanics. All these theories violate one or more principles which traditional philosophy considers as necessary conditions of science.

Obviously, these and similar developments in science gave rise to fierce opposition on the part of traditional philosophy. At first, its adherents made attempts to eliminate the theories just mentioned as scientifically invalid. As these attempts met with no response among men of science, philosophers had to change their tactics. They now declared that the theories in question, though perhaps useful in specialized scientific research, were entirely irrelevant from the philosophical point of view, and therefore had to be supplemented by a philosophical logic (Husserl, 1913), a philosophical theory of space (Becker, 1923; May, 1937), a philosophy of nature (Hartmann, 1950; Maritain, n.d.). These 'philosophical doctrines', however, bear a most striking resemblance to the views formerly adopted but now abandoned by men of science.

This situation naturally led to attempts to deal with the problems of philosophy in a strictly scientific manner, unbiassed by traditional opinions. These attempts initiated the development of scientific philosophy.

It should be noted that scientific philosophy does not indiscriminately reject the tenets of traditional philosophy or even of irrationalism. On the contrary, it considers it as one of its tasks to clarify the often obscure pronouncements of traditional philosophy and of irrationalism. Some of these must be dismissed as devoid of sense, others must be rejected as false; a number of these statements, however, can be accepted, though in most cases only after a thorough revision.

I would also like to emphasize that scientific philosophy does not necessarily have to take up a purely passive attitude towards science. On the contrary, it may influence the development of science by stimulating research and by exposing pseudo-scientific tendencies.

The most important methods applied in scientific philosophy are:

(1) Logical analysis, as propagated by Bertrand Russell and by the Vienna Circle.

(2) Significs, socio-psycho-linguistic or semiotic procedures (Mannoury, C. W. Morris, Hollitscher).

(3) Historical research (Couturat, Duhem, Enriques, Kelsen).

The power of the method of logical analysis is perhaps best demonstrated by Tarski's investigations into the Aristotelian or absolute notion of truth. The Ancients already observed that the unlimited application of this notion gives rise to a contradiction which is known as the Paradox of the Liar. Modern logic has given this paradox a form which shows that it cannot be dismissed as a mere fallacy.[1]

Tarski has established a definition of truth which does justice to the intention of Aristotle's definition without leading to contradictions. I will state this definition for a very simple formalized language.

The language consists of sentences S[2] which are constructed out of the connectives $^-$ and $\&$ and the atomic sentences $p_1, p_2, p_3, ..., p_n, ...,$ according to the following formation rules:

F1. For any n, the atomic sentence p_n is a sentence.

F2. If S is a sentence, then $\bar{S}$ is a sentence.

F3. If S and T are sentences, then $S\&T$ is a sentence.

F4. Nothing is a sentence if not by virtue of F1, F2 or F3.

Thus $\bar{p}_3 \& (p_5 \& \bar{p}_2)$ is a sentence, whereas $\overline{\& p_3 p_5}$ is not.

We now attribute a *meaning* to all sentences by means of the following interpretative rules:

I1. For any n, the atomic sentence p_n means that n is odd.

I2. If a sentence S means that A, then the sentence $\bar{S}$ means that not A.

I3. If the sentences S and T mean that A and B, respectively, then the sentence $S\&T$ means that both A and B.

In virtue of these interpretative rules, the sentence $\bar{p}_3 \& (p_5 \& \bar{p}_2)$ means that 3 is not odd and that 5 is odd and 2 is not odd.

We can now state which sentences are true and which are false by means of the following *truth definition*.

T1. For any n, the atomic sentence p_n is true if, and only if, n is odd.

T2. The sentence $\bar{S}$ is true if, and only if, the sentence S is not true.

T3. The sentence $S \& T$ is true if, and only if, the sentences S and T are both true.

T4. The sentence S is false if, and only if, it is not true.

According to these definitions, the sentence $\bar{p}_1 \& (p_5 \& \bar{p}_2)$ is false, whereas $\overline{\bar{p}_1 \& (p_5 \& \bar{p}_2)}$ is true.

From these definitions it follows that:

(i) no sentence can be both true and false,

(ii) every sentence is either true or false,

(iii) if a sentence S means that A, then S is true if, and only if, A, and S is false if, and only if, not A.

The conclusions (i) and (ii) obviously represent the principle of contradiction and the principle of the excluded third, applied to the sentences of our formalized language. Conclusion (iii) represents the Aristotelian definition of truth, as applied to these sentences.

However, our definition does not give rise to the Paradox of the Liar. As a matter of fact, truth and falsity are only defined with regard to sentences belonging to the formalized language. This language, however, contains no sentence analogous to the sentence pronounced by the liar, stating its own falsity. Such a sentence S would have to fulfil the following condition:

By virtue of the interpretative rules I1, I2 *and* I3, S *means that* S *is false.*

Conclusion (iii) now allows us to derive:

S is true if, and only if, S is false, and S is false if, and only if, S is not false.

However, it has been laid down in the interpretative rules (i)–(iii) that a sentence of our formalized language can only mean that certain numbers are odd or not odd. Thus a sentence meaning that a certain sentence is true or false does not occur in this language.

The definition of truth (and of a number of related notions, e.g., the notions of meaning, denotation, definition, consequence, synonymity) is only one of the fruits of Tarski's introduction of the *semantic method*. Other results are the exact interpretation of logical formulae, allowing a more satisfactory foundation of the formal laws of reasoning and a clearer insight into the non-formal presuppositions of symbolic logic and of formal axiomatics. In our example, for instance, we had to assume the existence of the series of the natural numbers as well as the validity of definition and proof by recursion; in more complicated cases we have to rely on even less elementary methods belonging to the theory of sets.

Semantics provides us with a "doctrine des vérités préalables" as postulated by Gonseth. I think it a pity that Gonseth does not so far accept the methods and the results of semantics. Only a few months ago (Gonseth, 1948) he repeated two objections which he had already made several years before (Gonseth, 1938, 1939) and which, in his opinion, had not been answered, namely: (1) semantics is based on naive realism, and (2) it is intended to give once for all a basis for all scientific activity. As a matter of fact, the first objection has been dealt with by Tarski himself (1944), while the second betrays a complete misunderstanding of the aims, the methods and the results of semantics. Therefore I would like to answer both objections together by pointing out that semantics in no way adopts a dogmatic point of view. It aims at clarifying the foundations of science, not at establishing an unshakable conviction. It achieves this by laying bare the hidden assumptions underlying symbolic logic and formal axiomatics. It does not claim to be able to justify these assumptions, but only shows that modern science takes them for granted.

A few remarks should be added about irrationalism. Though its origins lie in the past (Pascal, Rousseau, Kierkegaard), it derives its recent impetus in a large measure from the friction between traditional philosophy and contemporary science, which is of course detrimental to the rational attitude prevailing in traditional philosophy as well as to the prestige of science. Abandoning both the rational attitude and the contact with science, irrationalism succeeds in preserving the phraseology of traditional philosophy.

Recent work by Bertrand Russell (1947), K. R. Popper (1946) and Magdalena Aebi (1946) has shown the devastating influence of irrationalism in the social, cultural and political field. Modern totalitarianism derives from irrationalism, whereas democracy is traditionally connected with rationalism. Rationalism demands the freedom of discussion in all domains of human interest which only democracy is able to safeguard. Democracy and rationalism are indispensable conditions of human progress.

These last remarks should not be considered belonging to or deriving from scientific philosophy. They are based on a sociological reflection on the social conditions of scientific research. Nor should these observations be understood as lessening the value of the scientific research performed by those who do not enjoy the blessings of democracy. On the contrary, their merit is proportional to the hardships and the perils accompanying their work. However, it should be clearly stated that in the long run science and scientific philosophy can only prosper under democratic rule.

It could be objected that the propagation of erroneous doctrines can be more efficiently opposed by a totalitarian government. It must be admitted that in our days pseudo-sciences like astrology greatly benefit from the intellectual freedom obtaining in democracies. In my opinion, we should take this evil into the bargain. Indeed, the very progress of science is dependent on the toleration of error. Just like democracy, it has means of self-correction at its disposal which any totalitarianism lacks. When scientific information will be sufficiently wide-spread – which at present is nowhere the case – pseudo-science, at least in its cruder forms, will vanish on its own account.

REFERENCES

[1] Such an attempt was recently made by A. Koyré (1946, 1947a). The utter worthlessness of his arguments was quite convincingly demonstrated by Yehoshua Bar-Hillel (1948). Cp. Koyré (1947b).

[2] 'S', 'T', ... are variables taking the sentences of the formalized language as values. p_1, p_2, p_3, ... are the atomic sentences and $^-$ and $\&$ are the connectives which this formal system contains; i.e., the symbols 'p_1', 'p_2', 'p_3', ... occurring in the text are names for these atomic sentences and connectives; they are not identical with these atomic sentences or connectives themselves. Consequently, these atomic sentences and connectives are only mentioned, not used.

'A', 'B', ... should be considered as abbreviations of arbitrary sentences belonging to the language of elementary arithmetic, which is used as a metalanguage.

V

NIEUWENTYT'S SIGNIFICANCE FOR THE PHILOSOPHY OF SCIENCE

Though Nieuwentyt's scientific work is not extensive, it may be useful to give a brief characterization of his writings on purely scientific subjects. A few contributions to mathematics are known; the most important are those, presumably, in which Leibniz' foundation of the calculus was attacked.

A few years before his death Nieuwentyt published a bulky volume on *Het regt Gebruik der Werelt Beschouwingen, ter Overtuiginge van Ongodisten en Ongelovigen aangetoont* (The Right Use of Contemplating the Works of the Creator, Amsterdam 1716). It is mentioned in this context because it gives evidence of Nieuwentyt's thorough knowledge of the natural sciences of his time. However, the description of the universe which it contains was intended to convince "the *Atheists* of the wisdom, power, and goodness of their GOD, the venerable Creator and Ruler of the Universe; and *Infidels* who avow a God but by no means the authority of the Holy Writs, of the Supernatural Origin of the SCRIPTURE". The success of this book was enormous; of the Dutch edition at least seven printings are known (the seventh being of 1759). In addition, the book appeared in English (London 1718; 4th printing, London 1730), French (Amsterdam 1727), and German (Jena 1747). It was greatly admired by Rousseau and by Chateaubriand; the latter even gave a summary of it in his celebrated *Génie du Christianisme* (tome II, Livre V).

In my opinion, Nieuwentyt's book on *Gronden van Zekerheid of de regte Betoogwyse der Wiskundigen so in het Denkbeeldige, als in het Zakelijke: Ter Wederlegging van Spinosaas Denkbeeldig Samenstel: En Ter aanleiding van eene Sekere Sakelyke Wysbegeerte aangetoont* (Foundations of Certitude or the right Method of Mathematicians, in the Imaginary as well as in the Real: demonstrated in order to refute Spinoza's Imaginary System: and to introduce a Certain Real Philosophy, Amsterdam 1720) is still more interesting, even though the success of this book was less spectacular (third printing, 1739). Unlike Nieuwentyt's other writings, this work seems to have fallen into complete oblivion.

This lack of success is not difficult to explain. The title of the book suggests a tract of the edifying and polemical type which is so characteristic of Nieuwentyt's time. This description, however, only fits the second half of the work, and even so only with some qualifications; the first half is of a completely different nature, and presumably put off many prospective readers especially interested in books of a polemical and edifying nature. The following summary of parts I and II, which together occupy more than 200 pages, i.e., more than half of the book, will make this clear.

Part I deals with *pure* or, as Nieuwentyt mostly says, *imaginary* mathematics. This discipline is concerned with pure ideas, which Nieuwentyt also calls *entia rationis*. It is not necessary that such ideas be *real*, it is sufficient that they be possible. The notion of possibility must, of course, be explained, and it is worth-while to follow Nieuwentyt somewhat more closely here. The criterion of freedom from contradiction is mentioned, but not accepted. Nieuwentyt states that in pure mathematics an idea may be admissible although we do not know whether it could be real, and even if we do know that it could never be real (by way of illustration, Nieuwentyt mentions a magnitude of more than three dimensions). Furthermore, an idea may be admissible even though we know it to be contradictory, for otherwise, as Nieuwentyt rightly observes, proof by *reductio ad absurdum* would be impossible.

Consequently, theorems of pure mathematics are always conditional, that is, they consist of a hypothesis and a thesis. They may be true if the objects to which they refer do not exist. Interestingly, Nieuwentyt extends this view even to statements of a universal character, which strikes us as quite modern.

Nieuwentyt repeatedly points to the fact that a pure mathematician is not interested in questions regarding the origin of the ideas with which he is concerned, nor in the question whether these ideas are pure or real. Mathematical argument is then described as a process in which ideas are considered and compared and in which agreement or difference with respect to certain properties is 'experienced'. With a term which Nieuwentyt borrows from Descartes this kind of experience is characterized as *clear and distinct understanding*.

In accordance with tradition, Nieuwentyt adopts the absolute or

Aristotelian notion of truth as the truth concept, and the clear and distinct understanding as the truth criterion of pure mathematics. The conclusions of Part I are summed up in the heading of the last chapter:

Imaginary or Speculative Mathematics as a whole can teach nobody anything more than the Thoughts or mutual Relations of mere Ideas of the Ratiocinator, or mathematical consequences from these Ideas, which He or somebody else arbitrarily proposes to himself to be investigated.

Part II is devoted to applied mathematics. The purpose of this branch of mathematics is to find truths concerning *things*, that is, objects which really exist.

A basis is found in certain fundamental experiences, any appeal to hypotheses or to mere ideas being rejected. In this connection, Nieuwentyt refers to Newton's preference for the analytic over the synthetic method. It is permitted, however, to accept experiences of somebody else 'on trust'.

Starting from the fundamental experiences, then, we form ideas concerning things, which may be expressed by means of definitions, hypotheses, or axioms. Nieuwentyt agrees with an observation made by Keill, according to which things ought to be defined in terms of their properties. This is not to be construed as an anticipation of the method of 'implicit definition'; the remark is meant as an objection to definitions *per essentias & naturas*, a method which Nieuwentyt finds in Wolff.

The Real Ideas thus obtained are now divided into true and false ideas; the true ideas, on the one hand into perfect and imperfect ideas, and on the other hand into adequate and abstract ideas. Thus abstraction, according to Nieuwentyt, only occurs in applied, not in pure mathematics; this means a radical departure from traditional conceptions. Nieuwentyt observes that an abstract idea may be a perfect idea of certain properties of a thing and yet an imperfect idea of the 'whole' thing.

Nieuwentyt then makes an extremely important remark: *there is no difference whatever between real and imaginary argument.* This remark at once invalidates any attempts at completing formal logic by constructing a logic of content. It is well-known that numerous attempts have been made in this direction during the 19th century.

Having accepted the absolute notion of truth, we know how to ascertain the truth or falsehood of a given proposition. There are two questions to be answered, namely: (1) what does the proposition state, and (2) which are the properties of those particular things to which the proposition refers?

The connections between *argument* and *truth* are discussed in a rather prolix manner, a prolixity due to the absence of a solid logical basis. Nieuwentyt cannot reasonably be blamed for this; he should much rather be praised for avoiding inconsistencies. It should be noted that such logical methods as were available are aptly used. I restrict myself to rendering the main conclusions of Nieuwentyt's argument.

(i) If a real idea is perfect and adequate and if a deduction from this idea is necessary and mathematical, then the conclusion is true with respect to the whole thing;

(ii) if a mathematical conclusion is false with respect to the whole thing, then the real ideas involved are imperfect;

(iii) a conclusion from abstract ideas is conditionally true;

(iv) a conclusion from a false idea may be true;

(v) a conclusion from abstract ideas is true with respect to the whole thing if it is corroborated by experiments.

Nieuwentyt points out that one often has to start from ideas which are only approximately true. This does not affect the certainty of our conclusions, provided we avoid making extrapolations.

Nieuwentyt states a number of rules which one should keep in mind when searching for truths. A few of these rules will be quoted.

(iii) Even when all experiments confirm the conclusion based on certain suppositions, this is not sufficient for accepting these suppositions as true;

(iv) but if a conclusion obtained by a necessary and mathematical derivation is falsified by an experiment, then the suppositions are certainly false;

(v) in order to prove a proposition false it is not necessary to point out an error in its derivation (though this is often demanded by philosophers); it is sufficient to point out a false consequence.

This brings out the importance of rule

(ii) one single experiment proves both the truth of a certain conclusion and the falsity of its negation.

Rules (iii)–(v) are based on the theory of hypothetical syllogisms. Further, Nieuwentyt points out that even *reductio ad absurdum* only yields conditional truth. Therefore, the truth of conclusions can only *a posteriori* be extended to whole things. In 'real' mathematics, the criterion of truth is not based upon the 'clear and distinct understanding' but exclusively on *'experience and trust'*. This holds even for a proposition such as: *A Man who is able to hold and lift an Iron Bar of thirty pounds with his one Hand is also able with the same Hand to hold and lift a Bar of ten pounds of the same size.* In other words: 'real' mathematicians use their intelligence and their arguments not as *grounds* for 'real' truths, but merely as *tools*. This point of view is the exact opposite of the attitude of speculative philosophers such as Spinoza or Hegel.[1]

The ensuing introductory discussion of the ontological argument (discussed in more detail in the second half of the book) is outside the scope of the present essay. Nieuwentyt's observations on the notions of infinity and of probability, though interesting in many respects, must also be passed by in silence.

However, I wish to devote a few words to the final chapter of Part II, where Nieuwentyt sets out to explain how in mathematics the truth of conclusions obtained by valid inference may be ascertained, and where he defends the view that, essentially, veritable *mathematicians* do not differ from *logicians*.

A conclusion, then, is acceptable if:

(i) the inference is valid, and

(ii) the antecedent is true with respect to the whole thing.

An inference is valid if nothing is stated in the consequence which is not 'really' (though, perhaps, in different words) stated in the antecedent.

Nieuwentyt now goes on to point out that:

(I) By mere inference, however valid, no conclusion can be proved to be true, if not in the end either the immediate premisses or the next premisses or ... are proved to be true, *not by inference* but by *experience*; this holds for both pure and applied mathematics. And, conversely:

(II) A 'real' experience proves a given proposition to be true, not only if the experiment proves that very same proposition but also if it proves another proposition of which the first is a consequence.

From these considerations it appears that *veritable Logicians*, acknowledging the same rules as *Mathematicians*, differ from them only in the external form of the argument but not as regards the conclusive force of the methods of proof.

I must restrict myself to this summary of the first two Parts of Nieuwentyt's *Gronden van Zekerheid*, but it seems to me that we have already obtained a safe basis for a few preliminary conclusions, partly referring to Nieuwentyt's results and partly to his method.

As to Nieuwentyt's results, they do not nowadays strike us as spectacular. Still, in spite of many differences in terminology, his conceptions present a surprising degree of similarity to the philosophies of science developed during the past fifty years, and in particular to neo-positivism or logical empiricism.

(i) A sharp distinction is made between pure and applied mathematics;

(ii) in pure mathematics any ideas are admissible (formalism);

(iii) the conditional nature of the theorems of pure mathematics is stressed (cf. Russell);

(iv) pure mathematics is reduced to noting relations of similarity and distinction (formalism);

(v) pure mathematics "cannot teach us anything about reality" (neopositivism);

(vi) special attention is given to the problem of inter-subjectivity (neopositivism);

(vii) definitions ought to be given in terms of observable properties (positivism);

(viii) there is no difference between 'imaginary' and 'real' reasoning (the formal or tautological character of logic, emphasized by neo-positivism);

(ix) special attention is paid to approximative and statistical procedures;

(x) logical methods are used in the construction of a methodology of empirical science;

(xi) a radical empiricism is defended;

(xii) logic and mathematics are identified (logicism).

It seems to me that Nieuwentyt's views are nearer to contemporary conceptions than the doctrines of empiricists such as Locke or even Mill.

Still more strikingly modern, however, is Nieuwentyt's method, which in this connection ought to be compared, for instance, with the methods applied in Descartes' *Discours*, Pascal's *Esprit géométrique*, Leibniz' *Nouveaux essais*, and Mill's *System of Logic*. The last-mentioned works are discursive, normative and even speculative in their general approach, though not all to the same extent. Those parts of Nieuwentyt's *Gronden van Zekerheid*, however, which have been discussed here have a strictly analytical and descriptive character. The author does not lay down how investigations in the fields of pure and applied mathematics ought to be carried out but restricts himself to describing how such investigations are carried out in fact. His constructive contribution consists merely in pointing out the considerations on which the choice of a certain procedure is based; he does not, like Descartes, Leibniz, or Kant, indulge in speculations of an epistemological nature.

In accordance with this attitude, Nieuwentyt quotes numerous authorities in pure and applied mathematics[2] – he proves to be completely familiar with their work –, whereas only a few philosophers are mentioned.[3]

In my opinion, we should hail Nieuwentyt as one of the earliest workers – if not the very first – in the positive theory of science and in the critique of science, two disciplines which only since about a century have been cultivated more or less systematically. It is true that Nieuwentyt did not explore these fields for their own sake, but this hardly detracts from his outstanding merit. On the contrary, we are inclined to admire even more his consistency in keeping to his point of view, never giving way to the temptation to indulge in psychological, epistemological or metaphysical speculation.

Therefore I am very grateful to my friend Ph. H. Krijgsman who, a few years ago, handed me two impressive volumes, wondering if they might be of any value. I have made a provisional assessment[4]; much work remains to be done before a final verdict can be given. I hope it will be done, for Nieuwentyt certainly deserves it.

REFERENCES

[1] In the theses A to G on pp. 132-3 this point is explained once again, though in a more succinct manner.

[2] Namely: Alhazen, Apollonius, Archimedes, Barrow, J. Bernoulli, Tycho Brahe, Cassini, Clavius, Copernicus, Euclides, Flamsteed, Galilei, Gemma, Gregory, 's Gravesande, Halley, Hipparchus, la Hire, Huygens, Kepler, Leibniz, Mariotte, Newton, Proclus, Ptolemaeus, Fr. van Schoten, Stevin, Varignon, Wallis, and Johan de Witt.

[3] Keill, Raphson, Spinoza, Whiston, and Wolff.

[4] Cf. Beth (1959), pp. 39–41.

VI

SYMBOLIC LOGIC AS A CONTINUATION OF
TRADITIONAL FORMAL LOGIC

The aim of this essay is to call attention to the very close historical and systematical connections between traditional or 'Aristotelian' and modern or symbolic logic.

In order to appreciate these connections we shall have to characterize Aristotelian logic more precisely, and in particular discuss the question in what sense Aristotelian logic can be called formal. Let me observe, following Scholz (1931; cf. also the fundamental works Łukasiewicz, 1935 and 1951), that in every judgment and in every argument traditional logic distinguishes between the variable elements constituting the *content* and the constant elements constituting the *form*. Consider the judgment 'Socrates is mortal'. If the subject 'Socrates' is replaced by 'Julius Caesar' or 'Rembrandt' or 'Kant' and the predicate mortal by 'famous' or 'profound', the resulting judgment has a different content but the same form; this common form of the two judgments is expressed in the copula 'is'.

Now let us consider the classical inference

> All men are mortal
> Socrates is a man
> ―――――――――――――――
> ∴ Socrates is mortal

If we substitute 'Greek philosopher' for 'man', 'clever' for 'mortal', and 'Plato' for 'Socrates', we get

> All Greek philosophers are clever
> Plato is a Greek philosopher
> ―――――――――――――――――――――――
> ∴ Plato is clever

We see that the validity of the inference does not depend on its content but on its form. Some arguments, apparently, continue to be valid if we change their content but retain their form. This leads us to consider arguments abstracted from their content. Thus we get a discipline in which, not judgments and arguments, but judgment-forms and argument-forms are considered, and which we are quite justified in naming *formal logic*.

Formal logic must be clearly distinguished from any philosophical activity which, like Kant's transcendental logic or Hegel's dialectic, is concerned with the content of judgments and arguments as well. I shall restrict myself to a discussion of formal logic in this essay.

Aristotle's systematic treatment of formal logic is the oldest that has come down to us, and from Aristotle the greater part of what 19th-century logic textbooks and reference books present as formal logic is derived. In these textbooks and reference books, however, the exposition of the formal logic is almost invariably interspersed with discussions of topics from other disciplines, more or less related to logic – methodological discussions in Mill, epistemological and psychological ones in Schuppe and Sigwart, and metaphysical ones in Ueberweg (Mill, 1843; Schuppe, 1878; Sigwart, 1924; Ueberweg, 1888). If we leave these appendages and admixtures out of account, all that remains is a meager extract of the progress in formal logic made between Aristotle and Kant. About modal logic, which had already been developed considerably by Aristotle, hardly anything is to be found in 19th-century logic books. We do find in these books the theory of hypothetical and disjunctive syllogisms or argument-forms, deriving from Theophrastus and Eudemus; nothing at all however about propositional logic, which comprises this theory as a special case, although it was begun by Chrysippus and was considerably developed by Boethius, Abelard, and Peter of Spain.

Propositional logic is based on the observation that not only terms, but also judgments as a whole may be taken as the content of a judgment. Take, for example, the judgment: 'if it rains, then the sky is overcast'; here the judgments 'it rains' and 'the sky is overcast' are variable elements. If we replace them by the judgments: 'today is Wednesday' and 'it blows', the result is once more a judgment, namely 'if today is Wednesday, then it blows'.

Aristotle already made use of the now familiar method of specifying an argument-form, i.e., by writing *letters* in place of the variable elements in the linguistic rendering of the judgment. Thus a syllogism may for instance be formulated as follows (Aristotle, *Prior Analytics* A iv, 25^{b}32):

εἰ τὸ Α κατὰ παντὸς τοῦ Β καὶ τὸ Β	if in general *A* is predicated of *B*
κατὰ παντὸς τοῦ Γ, ἀνάγκη τὸ Α	and *B* of *C*, then necessarily *A* is
κατὰ παντὸς τοῦ Γ κατηγορεῖσθαι	in general predicated of *C*.

In this case the variable elements are *terms*. In propositional logic judgments can also be taken as variable elements. Now Boethius replaces such a judgment by one of the following expressions: '*a est*', '*b est*', '*c est*', '*d est*'; in this manner he is able to specify the form of a judgment or argument. Nowadays expressions of this kind are called propositional variables (for more particulars, see K. Dürr, 1938/39, 1951).

A fairly complicated argument-form occurring in Boethius is:

> *si, cum est a, est b, cum sit c, est d*;

which might be rendered[1] as follows:

> if from *a*, *b* follows, then from *c*, *d* follows.

In the notation of symbolic logic this becomes[2]:

$$(a \to b) \to (c \to d).$$

Modern languages employ syntactical devices to express what in modern symbolism is expressed by means of brackets (even, in some cases, simply by the arrangement of the symbols). Boethius' method is to use the synonyms *cum* and *si* alternatively, in accordance with certain rules. Here is an example of a judgment that has the form specified above: 'if from the fundamental principle of critical philosophy the impossibility of proving the existence of God follows, then from the possibility of rational theology the untenability of Kant's philosophy follows'.

We see that Latin is a little better suited than modern languages to specify logical forms. Modern symbolism, however, surpasses modern languages as well as Latin, especially where more complicated forms are concerned.

The principal feature that distinguishes modern symbolic logic from traditional formal logic is the much larger extent to which it makes use of expressly chosen symbols. Yet even in the earlier stages of the development of logic there is an obvious trend towards formalization, manifest, firstly, in the use of letters to indicate the positions of the variable elements and secondly, in the use of fixed, recurrent phrases. The constant elements in judgments and arguments, however, are usually specified by means of words in traditional formal logic.

It will have become clear that this distinction is only a matter of outward appearance. Symbolic logic is the crowning achievement of a trend towards formalization which is clearly perceptible already in Aristotle.

However, it is still frequently thought that as a result of its formalization logic has on the one hand become part of mathematics, and on the other hand is applicable only to the exact sciences. Designations like 'mathematical logic', 'algebraic logic' and the like have assisted in establishing this misconception. It is felt that as a result of formalization logic will have to be confined to the analysis of quantitative concepts and that it is incapable of dealing with the qualitative.

The answer would be that the rationale of the tendency towards formalization is the same in logic as in mathematics, namely the need for a concise and transparent vehicle of thought, and the inability of natural languages to provide one. Of course the formalization of logic has benefited from the example of mathematics. As a result there is a great outward similarity between the formulas of symbolic logic and those of algebra; but this is true of the older logical systems (such as Boole's) to a greater extent than of more recent systems. Moreover, symbolic logic was first applied in the analysis of mathematical arguments, not only because mathematics, for reasons which I shall not consider here, was in urgent need of such an analysis, but also because mathematics, with its comparatively transparent logical structure, was an obviously suitable subject for putting the methods of symbolic logic to the test.

Yet symbolic logic, like traditional formal logic, is in the first place a logic of qualities; it even attempts to reduce quantitative concepts to qualities. It claims to be applicable wherever there is reasoning, not in the sense of prescribing the various scientific disciplines certain forms of reasoning to the exclusion of other forms, but with the intention of specifying and discussing systematically the forms of reasoning current in the various sciences. This has always been the aim of formal logic, and it is difficult to see what fundamental objections could be made against it. Gradually other domains than mathematics are beginning to be analysed with logical methods, and it may be expected that the use of these methods will steadily increase.

The question may now be raised: if symbolic logic is nothing but the old, familiar formal logic, though supplemented at several points, adapted to more recent views, and phrased more effectively with the help of symbols that have been introduced expressly with this purpose, why then do logicians and others make so much of the difference between symbolic logic and traditional formal logic? Why then should logicians make such a fuss about the new insight allegedly gained by means of symbolic logic? And why do others, who appear to set much value on traditional formal logic, so strongly object to the ideas of modern logicians? Is the emphasis on the continuity in the development of logic not too one-sided and does this development not show a radical discontinuity somewhere?

In fact a discontinuity can be pointed out. It consists in the new orientation of logic originating with Leibniz, though partly based on ideas of Descartes. Alongside and even above traditional logic Descartes wanted to establish a new way of thinking which he named *mathesis universalis*. Its fundamental principles are embodied in the four, well-known methodological rules in the *Discours de la méthode* (Beth, 1957a). Leibniz, however, was the first to draw up the programme for a system of formal logic, which was to include Descartes' *mathesis universalis*.

Leibniz' programme becomes intelligible as soon as we form a clear idea of the requirements for formal logic: it should aim at *completeness* as to its subject matter and at *rigour* in its methods. The former means that formal logic ought to include *all* judgment-forms and argument-forms occurring in science. The latter implies that it must treat these judgment-forms and argument-forms systematically and that this treatment must bear the most critical examination.

According to Leibniz a system of formal logic meeting all these requirements presupposes three things, viz.: a *characteristica universalis*, a *calculus ratiocinator*, and an *ars combinatoria*.

By *characteristica universalis* we are to understand a *universal language of science*, in which all scientific reasoning and concept formation can be expressed. The *characteristica universalis* is not necessarily an expressly constructed system of symbols; there are, on the contrary, several possibilities. The *characteristica universalis* may be:

(1) a natural language, e.g., Latin;

(2) a partly formalized natural language; the formalization might for

instance consist in fixing the meanings of certain terms in contra-distinction to usage in ordinary language, which usage, since it depends on context, is never constant; or it might consist in the use of certain chosen fixed phrases – an example of this is provided by the language of elementary physics text-books;

(3) a partly formalized language as described under (2), but supplemented with expressly introduced symbols – this type is exemplified by the language of mathematics text-books;

(4) a completely formalized system, which may, however, contain some assimilated rudiments from ordinary language.

The *calculus ratiocinator* is a complete *system of argument-forms* providing, in a systematical form, all those operations which, given certain premisses, yield a conclusion, independently of the content of the statements involved. In other words, this calculus is a formal theory of deduction.

The *ars combinatoria* is a complete *system of definition-forms* and thus provides, in a systematical form, all those operations which, given certain concepts, yield new concepts, independently of the content of the concepts involved. Hence it may be considered as a formal theory of definition.

We already saw that the *characteristica universalis* is not necessarily a completely formalized system. We should now add, however, that a *calculus ratiocinator* and an *ars combinatoria* can only be formulated precisely when related to some formalized system. Therefore, if we want to have at our disposal not only a *characteristica universalis*, but also a *calculus ratiocinator* and an *ars combinatoria*, our scientific language will have to be formalized completely. Once this has been achieved, the systems of argument-forms and definition-forms may be specified quite as precisely as the calculating methods of arithmetic or algebra.

Our next task is to ascertain in how far the system of traditional logic meets Leibniz' requirements.

In formulating the laws of definition and reasoning, traditional formal logic has recourse to ordinary language. Thus the question arises: is ordinary language a *characteristica universalis*? This question must be answered in the affirmative in so far as the expressive resources of ordinary language are unlimited, so that it may be continually adapted to the expressive needs of science. From a logical point of view, however, ordinary language often has highly unsatisfactory ways of expresssing things. In the sentences 'John and Peter are rascals' and 'John and Peter are brothers', one and the same sentence-form expresses two different logical functions: in the second example a *relation* is expressed as if the statement involved the predication of a *property*. Not only the words of ordinary language are ambiguous, but, even worse, also the grammatical forms.

Another difficulty is that different *suppositions* of a term are often denoted by one and the same word.

All this implies that argument-forms cannot be described precisely as long as our instrument of reasoning is ordinary language. Consider the argument

> John is a rascal
> Peter is a rascal
> ―――――――――――――――――――
> ∴ John and Peter are rascals.

Here 'John', 'Peter', and 'rascal' are variable elements, so that we get the argument-form

> A is a P
> B is a P
> ―――――――――――――――
> ∴ A and B are P's.

However, if we now substitute 'John', 'Peter' and 'brother' for A, B and P, respectively, we get:

> John is a brother
> Peter is a brother
> ―――――――――――――――――――
> ∴ John and Peter are brothers

an inference which apparently is not valid.

It is because of difficulties like these that traditional logic has not succeeded in identifying and classifying all argument-forms occurring in science.[5,10]

Finally, the theory of definition of traditional logic is totally insufficient; as proper definitions it only recognizes those *per genus proximum et differentiam specificam*. Take for example the definition "two natural numbers will be called incommensurable if there is no natural number different from 1 that divides both these natural numbers". One might be tempted to consider the concept of 'number' as the *genus proximum*; in this one would be mistaken, however, since it is not a special *kind* of natural numbers which is defined but a *relation* between numbers. Apparently it is difficult to find a proper place for this definition in the traditional theory.[3]

Relations were the *bête noire* of traditional logic. Plato already struggled with them and wanted to conceive the relation of size as a kind of mixture of large and small – a first attempt to reduce a relation to inherent qualities. Leibniz was the first to appreciate the difficulties that were bound to be involved in such attempts; according to him, a relation is an *accidens* straddling, as it were, the gap between the one object and the other. But even Leibniz did not yet succeed in giving relations a suitable place in formal logic (Russell, 1900).

It would not be difficult to mention a great number of other cases where traditional logic is equally deficient. I think, however, we may confine ourselves to the examples already quoted.

Indeed the deficiencies of ancient and medieval logic had been realized long before Leibniz. This was reflected, on the one hand, in a criticism of formal logic as such, in an endeavour to advance science without any appeal to a formal theory of argument-forms (Bacon, Descartes), and, on the other hand, in attempts to make formal logic a really universal theory of argument-forms; attempts of this kind were made by Valla and Vives with their theory of modalities and by Ramus with his description of oblique conclusions.[4] As we have seen, an ambitious programme for further research in this direction was drawn up by Leibniz. The history of both systematic philosophy and formal logic has been decisively influenced by the circumstance that Leibniz' programme could not be realized at once. For the development of systematic philosophy now became disengaged from formal logic, which had proved to be incompetent, and the study of formal logic became a non-philosophical pursuit. In the works of the great systematic philosophers formal logic came to occupy a less and less important place, while the quality declined proportionately.[5]

After Leibniz two trends may be discerned in the history of formal logic. The first does not lead in the direction pointed out by Leibniz, and apparently considers the field of formal logic to have been more or less completely explored by his predecessors; this is the view taken by men like Kant, Fries, Herbart, Lotze, Ueberweg, and Sigwart. The second development on the contrary attempts to carry out Leibniz' programme; here the names of Lambert, Ploucquet, DeMorgan, Boole, Peirce, Schröder, Peano, Frege, and Russell should be mentioned. Bolzano's position is intermediate between these two.

It may be considered as an established fact that the second development in the history of formal logic has led to the realization of Leibniz' programme – to the extent that this programme was capable of being realized – and thus to the construction of a system of formal logic which, as far as we can judge at this moment, is tenable as well as complete. However, this should not result in a one-sided glorification of those scholars who have pursued this aim, or in an underestimation of the importance of other kinds of research in the discipline concerned.

It cannot be denied that Leibniz and his followers have often proclaimed quite extravagant ideas about the importance of formal logic, which may to a great extent explain the aversion to their methods that is frequently to be observed. Firstly, there was the conception of formal logic as an *ars inveniendi*, a method of finding the answers to all sorts of questions by a kind of mechanical computation. Stanley Jevons even succeeded in constructing a 'logical piano'; however, the output of this reasoning device was low.

In more recent research about the so-called decision problem (Gödel, 1931) it has been convincingly established that formal logic will never be able to supply such a method (though it should be kept in mind that this research has become possible only as a result of the modern development of logic). For the solution of all problems of any significance, *ingenuity* continues to be indispensable, even though formal logic may facilitate or support our intuition. The great importance, however, of formal logic is that it enables us to formulate precisely the solution to a problem once it has been found, and to test it rigorously.

A second erroneous view that has frequently been upheld by the followers of Leibniz is that mathematics is independent of any appeal to what is intuitively given. Even formal logic itself appears to require some intuitive data (Beth, 1937).

On the other hand the writings of non-formal logicians often contain valuable results in formal logic, even though these have frequently been arrived at by symbolic logic as well, and are mostly in need of test and revision. I shall give an example.

In the view of Herbart, categorical and hypothetical judgments differ in their linguistic forms only, and a categorical judgment does not imply existence. Ueberweg[6] does not agree with this. Now we should make a distinction between singular judgments like 'Socrates is a man', and universal judgments like 'all men are mortal'. The first judgment expresses a pure inherence (in our example, the inherence of being a man in Socrates) and from it we may infer that there exists a man, namely Socrates. The second judgment only manifests its logical structure if we formulate it as follows: 'for every being: if it is a man, then it is mortal'. This implies, firstly, that this judgment does not permit the conclusion that a man or a mortal being exists, so that the usual *conversio per accidens* of the second judgment into 'some mortal beings are men' is not permissible either (as indeed Ueberweg puts forward against Herbart); this means that the syllogisms DARAPTE, FELAPTON, BAMALIP and FESAPO will have to be dropped. And secondly, it implies that the argument we already discussed

> All men are mortal
> Socrates is a man
> _____
> ∴ Socrates is mortal

should be analysed as follows:

> For every being: if it is a man, then it is mortal
> _____
> ∴ If Socrates is a man, then Socrates is mortal
> Socrates is a man
> _____
> ∴ Socrates is mortal.

I.e., it rests upon the *dictum de omni et nullo* and upon the *modus ponens*.

If we should want to put Ueberweg in the right and retain the syllogisms just mentioned, we would have to interpret the universal judgment 'all men are mortal' as 'there are men, and for every being: if it is a man, it is mortal'. This would however involve serious complications, since every proof of a universal statement would have to be supplemented with a proof of existence.

Traditional formal logic was closely related to certain metaphysical systems, in particular the Aristotelian system. Does this apply to symbolic logic as well?

We know that the neo-positivists made use of symbolic logic in their endeavour to make science independent of metaphysical presuppositions. But this is a far cry from refuting certain metaphysical systems by means of symbolic logic or from demonstrating the impossibility of all metaphysics. Thus the question remains: can symbolic logic help us to criticize a metaphysical system or to devise one?

One of the main arguments of Kant's *Critique of Pure Reason* against dogmatic metaphysics is the fundamental impossibility of the *mathesis universalis* which, as we have seen, according to Descartes and Leibniz would have to serve as the foundation of metaphysics as well as of the other sciences. As a matter of fact the shortcomings of the philosophies of Descartes and Leibniz are to a large extent explained by the imperfections of the logical resources which they had at their disposal; in particular, as I have already pointed out, these resources did not include an adequate theory of *relations*. One consequence is Descartes' failure to account for the relations between *extensio* (extension) and *cogitatio* (thought), so that his anthropology suffers from a serious lacuna.

Leibniz assumes a plurality of substances, but he cannot imagine relations between them: the monads are 'windowless'. In order to make a meaningful world nevertheless possible and in particular to answer the anthropological question left unsolved by Descartes, Leibniz is compelled to resort to the *harmonia praestabilita* (pre-established harmony), an emergency measure like the occasionalism of the Cartesians.

Even Descartes and Leibniz, however, believe in the reality of the relations of identity and non-identity.[7] Spinoza is more consistent in this respect, since he allows only one substance, which has *extensio* and *cogitatio* as its attributes, though this is not to say that there are no logical difficulties in Spinoza's doctrine.

So Kant can hardly be blamed for qualifying the history of metaphysics as a "*blosses Herumtappen*" (a mere groping in the dark). The *Critique of Pure Reason* is meant as a discussion of the question, why the attempts of metaphysicians have failed and what could be done about it. From a methodological point of view the Critique culminates in the *Transzendentale Methodologie* (Transcendental Methodology), Chapter 1, section 1, which essentially is just a wholesale polemic against the idea of a *mathesis universalis* along the lines of Descartes and Leibniz; in this connection the distinction that Kant makes between philosophical and mathematical method is particularly relevant. (A definitive systematical and historical commentary on this part of the *Critique* and on the closely related pre-critical work *Untersuchung über die Deutlichkeit der Grundsätze der natürlichen Theologie und der Moral* (Inquiry into the Clarity of the Fundamental Principles of Natural Theology and of Morality) dating from 1764 does not, as far as I am aware, exist.)

However, this is the very part of Kant's work that can hardly hold its ground now that the programme of Leibniz has been carried out. It cannot be doubted that the construction of a complete system of formal logic in principle allows the formulation of a metaphysics, in the sense of an all-embracing, logico-deductive philosophical system.

A modern system of metaphysics on the basis of the new logical developments has been devised by A. N. Whitehead (1926, 1929, 1933; cf. also Devaux, 1930). Like the classical systems, his metaphysics claims to supply an all-embracing philosophy. In many ways this philosophy reminds us of Leibniz' monadology, though the acceptance of relations as real renders the makeshift of the *harmonia praestabilita* unnecessary; still a rational theology is certainly not lacking. It would be interesting to compare in more detail Whitehead's system with that of Leibniz or, for that matter, with another modern monadological system such as the metaphysics of Nietzsche, who like Whitehead differs from Leibniz in accepting the reality of relations (Baeumler, 1931, especially pp. 36 ff.).

Nevertheless, such a system does not escape Kant's second argument – it does not take the possibility of experience into account, but rather remains on the level of '*Begriffsdichtung*' (conceptual imagery), so that it must be judged by the subjective impression made on the reader, not by its objective scientific value. In spite of its logical systematization, such a system does not constitute real science.

The Dutch novelist and essayist Multatuli likened nature "in its *stupidity* in view of its *generality*" with a pair of plate-shears, which cuts through a book, a young girl or a sheet of tin with equal indifference. This metaphor could be applied to symbolic logic: with impartial equanimity it cuts through either metaphysical houses of cards or solid scientific edifices.

I now want to show that in spite of all this, symbolic logic poses problems displaying a close analogy with a classical metaphysical question, usually called the problem of universals. In my opinion these problems are real scientific problems, which arise as soon as we try to give a clear *interpretation* of symbolic logic.

I propose to consider the fundamental definitions of propositional logic, which I shall formulate in the simplest possible way. They are definitions of the terms *judgment-form of propositional logic* and *provable judgment-form of propositional logic*[8].

DEFINITION 1:

(a) The propositional variables $p, q, r, \ldots$ are judgment-forms (of propositional logic);

(b) if P is a judgment-form, then so is $\bar{P}$;

(c) if P and Q are judgment-forms, then so is $P \rightarrow Q$;

(d) the class of all judgment-forms is the smallest class fulfilling the conditions (a)–(c).

REMARK 1: For $\bar{P}$ read: *not P*, for $P \rightarrow Q$ read: *if P, then Q*.

REMARK 2: On account of definition 1(a)–(c), the following are examples of judgment-forms: $p, p \rightarrow q, \bar{r}, \bar{p} \rightarrow q, \bar{p} \rightarrow (\bar{q} \rightarrow r)$; on account of (d), the following are not judgment-forms (of propositional logic): $\varepsilon, \rightarrow (p\bar{\varepsilon})$, $\bar{\varepsilon} \rightarrow \varepsilon$.

DEFINITION 2:

(a) The judgment-forms

$$(p \rightarrow q) \rightarrow [(q \rightarrow r) \rightarrow (p \rightarrow r)]$$
$$(\bar{p} \rightarrow p) \rightarrow p$$
$$p \rightarrow (\bar{p} \rightarrow q)$$

are provable (Tarski, 1956);

(b) if the judgment-form $P(x)$ [containing the propositional variable x] is provable, then so is the judgment-form $P(Q)$ [obtained by replacing x, wherever it occurs in $P(x)$, by the judgment-form Q];

(c) if the judgment-forms $P \rightarrow Q$ and P are provable, then so is the judgment-form Q;

(d) the class of all provable judgment-forms is the smallest class fulfilling the conditions (a)–(c).

REMARK 1: In definition 2(c) we recognize the *modus ponens*.

REMARK 2: I have departed from ordinary usage in calling the three

54

basic formulas or axioms mentioned in definition 2(c) *provable*. The reason is that the definition may be formulated more simply in this way. It does not mean that these 'axioms' follow from other, even more fundamental formulas.

Definitions 1 and 2 are not themselves judgment-forms of propositional logic. They are not judgment-forms at all, but judgments; they are formulated in partly formalized English, supplemented with a few symbols. Starting from these definitions we can prove various theorems. These theorems state, for example, the independence and consistency of the axiomatization of propositional logic given in definition 2. These theorems, as well as the arguments constituting their proofs, are formulated in the same way as definitions 1 and 2. Like these definitions, they do not belong to the (*completely*) formalized propositional logic; they constitute a theory called the *metalogic* or *syntax* of propositional logic, which has the formalized propositional logic as its *object*.

The problems referred to a little while ago arise in connection with the interpretation of *this* theory. According to definition 1 there are infinitely many judgment-forms; (a) and (c) alone suffice to yield, for instance, the infinite sequence

$$p, p \to p, p \to (p \to p), p \to (p \to (p \to p)), \ldots$$

This definition thus asserts by implication that there must be infinitely many expressions denoting these judgment-forms; we have just written down some of these. If we had formulated the definition somewhat more precisely, the existence of infinitely many expressions would have been expressed in a more explicit manner.

We have now run into a difficulty. If the supplies of ink, pencil, paint and chalk, as well as human energy, are finite, how could there ever be an infinite number of expressions, i.e., of physical inscriptions? In view of this consideration the definitions given would rest upon an assumption which is materially wrong.

In an attempt to solve this difficulty it has been suggested that judgment-forms are *not* denoted by empirically given physical inscriptions, the number of which could never be more than finite, but by (non-material) inscription-forms, of which an infinite number is certainly conceivable. The question now arises what is the relation between the physical inscriptions and the inscription-forms. Various answers are possible.

(i) An inscription-form is a class of physical inscriptions all having the same form (Tarski, Carnap). According to this theory, however, there could not be more inscription-forms than inscriptions, so that the inscription-forms would also have to be finite in number. This answer thus leaves us where we started.[9]

(ii) We could assume that, in accordance with the first answer, our mind gets acquainted with a finite number of inscription-forms, but subsequently is able to construct for itself an unlimited number of new inscription-forms.

(iii) We could suppose that our mind is able to construct the inscription-forms quite independently, without any preliminary abstraction, and that the physical inscriptions are then written down in accordance with these inscription-forms.

If we accept the second or the third answer, the inscription-forms constitute a potential infinity; therefore we can only talk about them intuitionistically, following the method applied by Hilbert and his school in their 'metamathematical' research.

(iv) We could, finally, adopt the view that the inscription-forms exist independently of the physical inscriptions and independently of our mind, and that the physical inscriptions as well as our mental representations of inscriptions and inscription-forms presuppose these inscription-forms as given. This not very plausible view is the only one by which the non-constructive proof-methods adopted particularly by the Polish school of metalogic can be justified.

It is intrinsically impossible to 'demonstrate', whether by analysis or empirically, the correctness of one of these answers. This will lead neo-positivists to consider the question as scientifically meaningless, i.e., as a pseudo-problem. Yet anyone who wishes to take a well-considered view of the scientific problems of formal logic will have to take sides in this question; if not, his work will lack the foundation of a suitable interpretation. This choice of a position is not, however, a matter of reasoning or of empirical research, but of temperament.

In the very animated discussion following the reading of this paper a number of fundamental questions were raised which I could only answer provisionally. I should like to discuss them somewhat more thoroughly here.

It appeared that the argument

> John is a rascal
> Peter is a rascal
>
> ∴ John and Peter are rascals

is acceptable, whereas the argument

> John is a brother
> Peter is a brother
>
> ∴ John and Peter are brothers

is not. In this connection the question was raised: do we not, in judging these arguments, take the content of premisses and conclusion into account and is not it the content that settles the matter?

To this question I would now give the following answer. If we formulate the two arguments in the above manner we soon come to realize that they do not have the same logical form; next we see that the first argument is valid though the second is not. This calls forth further investigations, eventually leading to a more precise characterization of the various logical forms.

In general the logical form of a judgment does not appear unambiguously from its formulation in ordinary language. Often the logical form only becomes apparent when the judgment occurs in arguments. Now the point of logical symbolism is to allow a formulation of judgments which displays the logical form unambiguously. As long as this end has not been attained, cases like the one discussed will occur.

Only after many unsuccessful attempts has symbolic logic succeeded in achieving a complete adjustment of its formulations to logical form. Thus the distinction between '*appartenance*' (appertaining) and *inclusion* has only been made in comparatively recent times. We find this distinction in Peano, but it does not yet occur in Schröder's great work (1890–1905). According to Peano, the judgments 'Socrates is a man' and 'man is mortal' represent two different logical forms. This has already been explained above, and also appears from a comparison between the following two arguments[10]:

Socrates is a man	An apple is sour
Man is mortal	Sour is a taste
∴ Socrates is mortal	∴ An apple is a taste

A second important question that was raised during the discussion concerned the field of application of symbolic logic. Does it for example apply to value judgments? Before we try to answer this question it should first be emphasized that the view that in such cases we should (or could) revert to traditional logic is untenable, and is based upon a mistaken conception of the relation between the two. Traditional formal logic no longer leads an existence of its own, alongside symbolic logic, but is part and parcel of it.

An answer to the question is suggested when we think of the origin of formal logic: it arose out of the needs of argumentation. In the oldest thinkers we never find a deductively coherent argumentation[11]; they restrict themselves to a dogmatic exposition of their views. Formal reasoning only occurs when the need is felt to refute an opponent or to defend one's own views against the views of others (Zeno of Elea, the Sophists). It lends a suitable outward form to the inner coherence of thought.

Thus the question whether the methods of symbolic logic apply to value judgments is reduced to the question whether value judgments can be the subject of a meaningful argument. This question cannot be decided on the basis of logical considerations alone, though it seems evident to me that an affirmative answer will have to be a postulate of any scientific theory of values.[12]

Our conclusion is, then, that a scientific theory of values is feasible only if value judgments are open to logical analysis. This is not to say, however, that the logical forms occurring in a theory of values would have to be the same as those occurring in other sciences. It is quite conceivable that for value theory there are specific judgment-forms and argument-forms.

Whereas pure symbolic logic and the logical analysis of mathematical theories have assumed quite impressive proportions, the application of symbolic logic to other scientific disciplines is still in its initial stages, even though some important contributions have already been made; for example the discussion of the concept of *type* by Hempel and Oppenheim, of the concept of *Gestalt* by Grelling and Oppenheim, and of quantum-theoretical logic by von Neumann and Strauss (Hempel and Oppenheim, 1936; Grelling and Oppenheim, 1938; von Neumann, 1932; Strauss, 1938).

Rather than summing up these investigations, however, I should like to say something myself about the logical form of ethical imperatives. In the interpretation which Grelling has given of imperatives they display a formal analogy with modal judgments; imperatives as well as modal judgments involve forms of reasoning that differ from those of 'assertory' propositional logic (i.e., as discussed above).[13]

In Carnap's interpretation of modal judgments, on the other hand, these judgments can be dealt with by the syntax of 'assertory' logic. Special modal forms of reasoning will then be unnecessary: the theorems of modal logic can be derived within syntax by means of 'assertory' forms of reasoning.

With respect to ethical imperatives the situation is similar. For the imperative 'one should keep a promise' may also be formulated as 'a decent person keeps a promise'. Judgments like these can be justified by deriving them formally from a *definition* of the form 'a decent person is somebody who behaves so and so', which either simply enumerates the kinds of behaviour manifested by decent persons or, preferably, gives a general characterization of such behaviour. For terms like 'good citizen', 'criminal', etc., similar definitions could be given. From these definitions we could then formally derive judgments like 'whoever commits a murder is a criminal', 'no criminal is a decent person', 'all decent persons are good citizens', and so on. If we should call 'ethics' the system of these definitions and of the statements that are logically entailed by them, we would have provided ethics with a purely rational foundation, independent of any appeal to experience.

There remains, nevertheless, an extensive field for empirical research. In the first place it should be examined in how far the definitions given are in accordance with customary linguistic usage. Furthermore, the question remains why individuals try to be a decent person, a good citizen, etc., or at least try to pass for one, and why the attitudes of people and groups toward individuals depend on whether or not they are decent persons, good citizens, etc.

REFERENCES

[1] At the meeting where the paper which is the original version of this essay was read, it was concluded that the alternative use of the indicative '*est*' and the subjunctive '*sit*' has no special significance in scholastic Latin.

[2] Cf. Theophrastus' formulation of the first hypothetical syllogism:

εἰ τὸ A, τὸ B; εἰ τὸ B, τὸ Γ; εἰ ἄρα   If *A*, then *B*; if *B*, then *C*; therefore, if
τὸ A, τὸ Γ.   *A*, then *C*.

[3] Strictly speaking we could consider this definition as *per genus proximum et differentiam specificam*, by taking the term 'pair of numbers' as the *genus proximum*. But then we would have to define the term 'pair of numbers', so that the difficulty returns at once.

[4] An example of such a conclusion is:

> The square of an even number is an even number
> 6 is an even number
> ———————————————————————
> ∴ The square of 6 is an even number

This is not an argument in the mood BARBARA, for then the minor would have to be: 'The square of 6 is the square of an even number'. But the transition from the premiss '6 is an even number' to this minor cannot be justified in traditional formal logic, since it has not at its disposal a *theory of descriptions*. A first commencement of such a theory can be found in John Stuart Mill, *A System of Logic* (Book i, Chapter ii, section 5), in his characterization of connotative individual names.

[5] For example, the quality of the *Logik* that was edited by Jäsche and published under Kant's name and with his authorization, is strikingly low. Thus *modus ponens* and *modus tollens* are introduced already in section 26, without the slightest systematic connection (though it appears from the notes left by Kant that he did not commit this blunder). J.G.C.C. Kiesewetter's *Grundriss einer reinen allgemeinen Logik nach Kantischen Grundsätzen* (Outline of a Pure General Logic on Kantian Principles) is on a much higher level. – It is sufficiently known that Kant's views on formal logic were very conservative (cf. *Critique of Pure Reason*, 2nd ed., p. viii). Yet now and then one comes across original and profound ideas on logic in his works. Thus in Book 2, Chapter 3, Section 4 of the *Transzendentale Dialektik* (Transcendental Dialectic), i.e., pp. 620ff. of the 2nd ed. of the *Critique*, he observes that existence cannot be a predicate. This was also the view held by the Scholastics, as appears from the fact that, following Avicenna, they distinguished between existence and essence. – The little progress that was at first made with the realization of Leibniz' programme can be partly accounted for by the authority of Kant, who thought this programme quite unfeasible. The lack of a suitable system of symbols, referred to at the meeting, does not, I think, suffice as an explanation. Round about 1800 mathematical notation was already strongly developed, while the achievements of Boethius and Abelard in propositional logic show how much one can do even without many symbols. – Even in our days Kant's authority is detrimental to symbolic logic.

[6] On this controversy, see Lange (1877), pp. 99–100; Trendelenburg (1862), vol. 2, pp. 248 ff.; Ueberweg (1888), pp. 162–163.

[7] Even in the more primitive stages of thought (magical, mythical, mystical, emotional thought) it seems that the identity relation does not meet with quite as much repugnance as other relations. This might account for the fact that in these stages of thought all relations were reduced to an identity relation. Even in philosophy various symptoms of this tendency are still perceptible: for example, in the identity philosophies in various periods of the history of philosophy (Parmenides, Spinoza, Schelling) and in the view that only judgments of identity are admissible (Stilpo, Antisthenes, Lotze).

[8] In this concise formulation of the principles of propositional logic I have not aimed

at ultimate precision. Thus I should, strictly speaking, have introduced special *names* for the signs by means of which propositional logic is formulated. Cf. pp. 31–2.

[9] As far as I know this problem was first formulated by A. Tarski (Tarski, 1956; cf. also Carnap, 1934).

[10] In Peano's notation, appertaining (i.e., the pure inherence of the predicate in the individual substance) is indicated by 'ε', while inclusion (i.e., the subsumption of a general concept under a broader concept) is indicated by $\subseteq$. The forms of the two arguments may now be represented as follows:

$$S\varepsilon M \qquad\qquad S\varepsilon M$$
$$M \subseteq P \qquad\qquad M\varepsilon P$$
$$\therefore\ S\varepsilon P \qquad\qquad \therefore\ S\varepsilon P$$

The first argument-form is valid, while the second is not. The following argument-form is also valid:

$$S \subseteq M$$
$$M \subseteq P$$
$$\therefore\ S \subseteq P$$

In traditional logic the judgments containing an '*appartenance*' and an *inclusion* were distinguished as *singular* and *general*, respectively, but this distinction was thought to be of no consequence for the evaluation of arguments. Cf. Kant, *Critique of Pure Reason*, 1st ed., p. 71; Ueberweg (1888), pp. 170–1. Burkamp (1932, pp. 81–2) does not accept the distinction of appertaining and inclusion, but his counter-arguments tell strongly in favour of the distinction!

[11] Compare, for instance, the ways in which Parmenides and Plato set forth their views. – About the origin of formal logic, cf. Kapp (1942).

[12] During the discussion something like the following, not uncommon, situation was brought up. *A* defends, as against *B*, the aesthetic value of a painting by pointing out a number of peculiarities of this painting. *B* then shows a second painting manifesting these same peculiarities, though obviously possessing no aesthetic value at all. Has *B* now refuted *A*'s arguments, or could *A* reply: the beauty of my painting resides in the very peculiarities that render yours worthless? By accepting this reply we should indeed paralyse any discussion from the outset. – In this connection I might mention Gaunilo's striking argument against the ontological proof of the existence of God. If somebody were to reply that you cannot argue about God in the same way that you can argue about an island, then there is no use in going on with the discussion; you may just as well simply establish the existence of God as a dogma. Cf. Beth (1955, 1958b).

[13] A similar view is already to be found in Valla (it is discussed in Lange, 1877, pp. 30ff.). For example, Valla puts the '*honestum est*' in '*honestum est pro patria pugnare*' (it is honourable to fight for your country) on a par with modal operators. From the view-point of formal logic, '*oportet*' (one ought to) is of course completely equivalent to '*honestum est*'. Cp. Grelling (1939).

VII

SOME REFLECTIONS ON CAUSALITY

1. I should like to introduce these general reflections on causality with the following provisional formulation of the causality principle:

(i) perceived or assumed events can be arranged in chains that are continuously linked in space and time;

(ii) these chains can be analysed into fragments each having specific structures.

The causality resides in the continuity of the chain as well as in the specific structure of the individual links. The above formulation is only intended to serve as a starting-point. I shall not discuss its adequacy, nor do I wish, just now, to go into the question of its epistemological status: *a priori* principle, postulate, law of nature or convention? A discussion of the question whether the principle is true and of its significance for science will also be postponed for a while.

My first concern is to show that there is much diversity as to the role which the concept and principle of causality play in different domains and at different levels of human thought. This explains why the answers to the questions indicated above are so strongly divergent, and it also suggests that a simple and unqualified solution of the problems of causality will not be possible. Very often the diversity of meaning of the concept of causality is not sufficiently realized.

2. The most primitive level is no doubt direct visual apperception of causal relationships as experimentally ascertained by Michotte (Michotte, 1946). There is a strong analogy between these results and the *Gestalten* that had appeared in the theory of visual perception some time before. The perception and recognition of these forms does not depend on their analysis into more basic elements. It seems to me that these psychological results are also of great significance for the theory of knowledge, even though they do not by themselves provide a complete solution of epistemological problems.

Results like those of Michotte may explain how it is possible that people with a minimal intellectual training may still, for example, become expert billiard-players. Apparently no more is required here than the causality that is given in perception.

Now such people may also learn to play chess very well. This might seem a far cry from billiards, since on reflection we see a great difference between the two cases. Taking a piece on the chess-board involves a *rule*, i.e. a *convention*, whereas the collisions of billiard-balls are a matter of 'real' causality. Our primitive subject, however, will probably not notice this difference at all.

Indeed the contrast between the two cases should not be exaggerated. In calculating the effect of a stroke at billiards not only the laws of mechanics are taken into account, but also certain rules of the game. Naturally these two kinds of causality will be inextricably linked with each other in the mind of our primitive subject, although his skill may be none the less for all that.

For numerous other kinds of performance it suffices to make use of the causality that is given in perception. Further examples are the activities of a cook and of a boxer. This is not to say that the intellect plays no role whatsoever in these and similar activities. At this level, however, the activity of the intellect is rather synthetic; it confines itself to constructing in anticipation causal chains that are given in perception.

3. The need for analytical reflection arises, however, as soon as we have to judge certain actions morally or juridically. Here *guilt* and *cause* are always closely linked, although in this connection interest is mainly centered upon guilt. Indeed historically, the concept of cause arose from the concept of guilt, as will soon become apparent. At this level there is no more occasion than there was in the first case to distinguish between conventional 'rules of the game' and 'real' causality. Guilt arises from wilful interference with an existing and generally valued order; in this connection it does not matter in the least whether this order is 'natural' or 'conventional'.

This analytical reflection in connection with the concept of cause, with a strongly marked moral or juridical background, dates back to very ancient times. It is manifested in a very curious *sociomorphous cosmology*, which was probably first investigated by Kelsen (Kelsen, 1941). Since the later development of the concept of cause has to a great extent been determined by this cosmology, a short discussion of it ought not to be omitted.

4. The universe is viewed as one great society, comprising not only mankind, but also the whole of organic and inorganic nature and even the gods. This cosmic society is governed by an order which derives from a divine law; all human laws, for instance, are valid by virtue of the divine law. This divine law implies

(i) a rule for the normal course of things;

(ii) a rule which imposes a certain compensation for every instance of interference with the normal course of things.

This curious view was still very much alive in the last century, for example, in Ralph Waldo Emerson's essay *Compensation*. Rather than dwelling on Emerson's ideas, however, I should like to give a very brief survey of the spread of sociomorphous cosmology in Antiquity.

5. Kelsen himself illustrates his exposition with examples taken from primitive imagination and from Greek religion and philosophy. Personally I do not think the primitive examples quite so convincing. We shall see that sociomorphous cosmology was very wide-spread in the civilizations of Antiquity; for this reason I seriously doubt the originality of these same or similar ideas in younger primitive peoples.

The Greek examples, however, are quite conclusive in my opinion, and I flatter myself that I have succeeded in supplementing the material gathered by Kelsen (Beth, 1952-53). Some striking samples must suffice here.

Anaximander:

And the source of coming-to-be for existing things is that into which their destruction, too, necessarily happens; for they pay penalty and retribution to each other for their injustice according to the order of Time (Diels-Kranz 12 A 9, translation after Kirk and Raven).

Heraclitus:

For all the laws of men are nourished by the one divine law (Diels-Kranz 22 B 114, translation after Kirk and Raven). – That which is opposite comes together and from different tones comes the fairest harmony (Diels-Kranz 22 B 8). – Sun will not overstep its measures; otherwise the Erinyes, ministers of Justice, will find him out (Diels-Kranz 22 B 94, translation by Kirk and Raven).

Plato:

Now all these become causes of disease when the blood is not replenished in a natural manner by food and drink but gains bulk from opposite sources in violation of the laws of nature (Timaeus 83 E, Jowett's translation).

Zeno of Citium:

Matter is moved by fate, which never changes and may be called either providence or nature (Aetius i, 27, 5; Diels, *Doxographi Graeci* 322 b 9).

Chrysippus:

This cosmos is a large scale state and is governed by one single constitution and by one law. And there is a law of nature prescribing what ought to be done and forbidding what ought to be left undone. – Our natures are parts of that nature which constitutes the whole, ... the universal law (Philo, *De Joseph*, volume 2, ed. Mangoldt, page 46).

Cicero:

Pythagoras and Empedocles declare that there is one and the same legal order for all living beings, and they proclaim that inexpiable penalties await those who have used violence against living beings (*De re publica*, Book iii, Chapter 11).

These quotations may have given a more lively picture of some of the main points of sociomorphous cosmology; the import of the cosmic law is that it maintains the harmony of the cosmos. I should also remind the reader of the importance of this moral or juridical causality for Greek tragedy. And finally I wish to mention in passing that Gregory Vlastos, apparently independently of Kelsen, reached quite similar results (Vlastos, 1947; cf. Heinimann, 1945).

6. Going back many centuries now, I first wish to mention the occurrence of sociomorphous cosmology in Egypt. The Egyptian way of thinking in this respect has been described very lucidly by Brede Kristensen:

To the modern mind cosmical law (law of nature) and ethical law are quite different things. The Ancients, however – not just the Egyptians, but most peoples of Antiquity, including the Greeks – had a different conception of these matters. They were familiar with the notion of an all-embracing order, determining the course of human lives as well as the course of nature, a universal law in which ethical and cosmic decrees lie closely together. *Ma-a-t* means the path of life, the order of eternal life, resurrection in nature and of man. Its natural side is manifested in the regular course of things. The eternal life of the sun-god is the order of *Ma-a-t*. Above all, however, it was thought to be at work in terrestrial life, which is eternal under all change. ...

Thus ethical law, which was obeyed by the dead, is here considered to be intimately related with cosmic law, or with order in nature. Its justice is at the same time the law of the sun-god; ethical law and the law of nature shade off into each other (Kristensen, 1926, pp. 67–75.)

All this has been more fully worked out by the Dutch historian of religion Bleeker, from whose thesis I quote the following short passage relating to the god Thot:

Properly speaking his task is to maintain the order of the universe. ... His functions are recounted in a hymn from which the following is quoted: "the lawgiver in heaven and on earth, who sees to it that the gods do not overstep the limits that have been set to their power, that each guild fulfils its duties, that countries know their boundaries, and the soil what is appropriate to it" (Bleeker, 1929, pp. 65–6; cf. also pp. 81–5).

7. Sociomorphous cosmology in Mesopotamia has been elaborately dealt with by Jacobson ('Mesopotamia', in Frankfort, 1946) and by David (David, 1949; and 'Ordre, destin, personne', in *Proceedings*, 1949). I must restrict myself to quoting a summarizing passage from the first author:

The Mesopotamian universe did not, like ours, show a fundamental bipartition into animate and inanimate, living and dead, matter. Nor had it different levels of reality: anything that could be felt, experienced, or thought had thereby established its existence, was part of the cosmos. In the Mesopotamian universe, therefore, everything, whether living being, thing, or abstract object – every stone, every tree, every notion – had a will and character of its own.

World order, the regularity and system observable in the universe, could accordingly – in a universe made up exclusively of individuals – be conceived of in only one fashion: as an order of wills. The universe as an organized whole was a society, a state.

8. About China details may be found in a recent study by the biochemist and sinologist Needham (Needham, 1951); his exposition fits in quite well with the picture that has now emerged. Sociomorphous cosmology, it is safe to say, is common to all the main civilizations of Antiquity and the Orient. It accounts for a variety of ideas which have come to appear strange to us, although its influence may be felt even in our days, in so far as, e.g., *astrology* and the idea of *natural law* are still flourishing. In

my view it should be qualified as a great conception, not as a product of mere 'primitive', 'a-causal' or 'emotional' thought. Obviously, however, the conception of causality which underlies sociomorphous cosmology could not but be a serious drag on the early development of modern science. This development became possible only when the power of sociomorphous cosmology had dwindled.

9. We have seen that the typical and leading representatives of Greek thought remained faithful to sociomorphous cosmology. This even applies to Democritus, although according to modern views neither his atomism nor his mechanistic conception of causality can be reconciled with sociomorphous cosmology. This was not, however, the view of the Ancients, and accordingly such few opponents of sociomorphous cosmology as there were – Strato of Lampsacus and Epicurus – have not been able to accept Democritus' mechanistic conception of the universe. As is well known, they modified this conception by introducing an indeterministic element, with the express purpose of robbing causality of its divine lustre.[1]

Matters became different, however, in later times. I would suggest that this is mainly due to the acceptance of the Christian conception of God. The moral or juridical notion of cause now came to be embodied in the idea of finality, whereas mechanical causality became an important factor in the development of modern natural science and thus a powerful weapon in the fight against sociomorphous cosmology. The great importance of mechanical causality for what is nowadays called *classical science* is, I presume, universally realized.

10. It is equally well known that, in the most recent physical theories, mechanical causality and the determinism indissolubly bound up with it have come to play a considerably less important role. Since a discussion of this point would require much space and would inevitably be rather technical, I shall not pursue it here.

However, it is not exclusively on a theoretical plane that the notion of causality is important. It plays a no less important part at the level of observation and experiment, which I had in mind more particularly when I gave my provisional formulation of the causality principle. Suppose we want to investigate a certain physiological process. Now we could proceed

67

as follows: we let the process take place in a number of laboratory animals. The process is however interrupted in all animals, though at a different point in each case. Thus we get a number of experimental results, which have to be combined in order to yield the outcome we wanted; it is here that we make use of the causality principle. For we assume that the process under investigation may be considered as a causal chain. In each of the animals we may observe one of the links. This fragment, we go on to assume, has a typical structure; i.e., it occurs invariably in *all* animals. Hence we mentally link the observed fragments and thus reconstruct the process under consideration as a causal chain.

If, in dealing with epistemological problems in connection with the causality principle, we focus our attention in the first place on the level here considered, the problems we have to face may be difficult, but they do not seem unsolvable to me. In view of the fact that there exists a spontaneous sensory apperception of the simplest causal connections, there is no objection to crediting the human intellect with a certain inclination to construct longer and even more complicated causal chains; at the experimental level this inclination would quite naturally lead us to devise experiments like the one just discussed. This does not imply, however, that the results obtained by means of such experiments will always be perfectly reliable; it does not even imply that these experiments will always lead to unambiguous results. Considered in this way, the causality principle is not an *a priori* principle; but neither is it a law of nature, and it certainly is not a convention. It would perhaps be wisest to say that the causality principle brings out, in a manner co-determined by historical factors, a universal human tendency to construct a far-reaching and if possible all-embracing causal structure on the basis of spontaneously apperceived causal and other connections, a structure which enables us to understand the universe in which we live as a cosmos.

REFERENCE

[1] Cf. Cicero, *De natura deorum* i (25, 69) and ii (16, 44).

VIII

SCIENCE A ROAD TO WISDOM

A book dealing with a number of problems in scientific philosophy would be incomplete if it failed to discuss the question in how far scientific philosophy can contribute to the attainment of the ultimate aim of all philosophical activity, namely, wisdom. The need to deal with this question becomes particularly pressing because frequently a view is held which raises doubt on this very point. Therefore the principal object of the following reflections will be to place scientific philosophy in a more favourable light in this respect. They should not be taken as a dismissal of differing forms of philosophical activity.[1]

In his article 'Het dal der na-oorlogse filosofie' (The Valley of Post-War Philosophy, 1958) H. J. Pos discusses the 'scientism' of contemporary philosophy, which in his view leads to relativity and self-estrangement. However, he hopes to see a "new classicism which takes up the thread that runs from Aristotle to Hegel" in the future.

Another Dutch philosopher, A. G. M. van Melsen, considers this scientism only apparent. According to him, even the philosophy of the exact sciences, which reflects on the presuppositions of these sciences, arrives at the "surprising conclusion..., that these presuppositions are not the fruits of science itself, but *pre*suppositions in the proper sense, derived from pre-scientific experience" (Van Melsen, 1954). From this it is inferred "that the development of natural science has no direct relevance for philosophical problems". It does have indirect relevance in so far as "the recent development of natural science has revealed the faulty and outdated formulations of former philosophical views". Thus we may "be confident that, in spite of the enormous gain in insight brought by philosophy of science, there is no question of a real conflict between that which a study of the great systems of traditional philosophy and of the great philosophers shows to be the key problems of philosophy, and that which may be gathered from philosophical reflection on science".

Logic is mentioned by Pos as a concrete instance of the scientism prevailing in modern philosophy. "Inspired by foundational research in mathematics, it has become an important branch of science. As a result,

69

the gap between the study of logic as pursued by the intellect, i.e. by science, and dialectical logic in the spirit of Hegel has widened; on the other hand there have been important renovations since Aristotle and the Stoa, the founders of logic." The same view is worded somewhat differently by Van Melsen: "It cannot be denied that much of what is still included among the topics of philosophy of science will soon have become positive science, or indeed has already become so. Symbolic logic is a case in point: for the greater part it must already be considered as belonging to positive science." Both authors appear to feel a sharp distinction between philosophy and (positive) science.

In my view Pos and Van Melsen fail to appreciate the historical and systematical connections between science and philosophy, as well as the philosophical import of recent scientific developments.

The vast territory of contemporary philosophy may be divided into three principal domains (cf. p. 29): (i) traditional systematical philosophy, (ii) modern irrationalism, and (iii) scientific philosophy, with which we shall be mainly concerned here.

The problems typical of all scientific philosophy have arisen in connection with the development of the exact sciences in the course of the last 100 or 150 years. This development has not only greatly increased our knowledge but also has led to radical changes and to the abandonment of views that for centuries had been considered self-evident (cf. p. 76). The repudiation of a variety of traditional views is the one common factor in the otherwise strongly divergent trends in scientific philosophy, some of which are: analytical philosophy, cantorism, dialectics, formalism, intuitionism, logicism, neopositivism, nominalism, significs, and philosophy of science (some of these trends are concerned with only one scientific discipline, but others extend beyond the domain of the exact sciences). Borrowing a witty phrase from Bachelard (1949) we might characterize scientific philosophy as a "*philosophie du non*".

In the same sense Gonseth, another member of the *Dialectica*-group, speaks of a "*philosophie ouverte*".[2] A very felicitous formulation is due to P. Bernays, who is also among the leading representatives of this group:

Indeed a philosophical reorientation is required. It is often thought that either we have to assume an absolute evidence or else we must give up all hope that any insight is to be gained by means of science. Instead of adopting this 'all-or-nothing' attitude it would be more appropriate to conceive of evidence as something which we acquire. We acquire evidences rather like we learn to walk or like young birds learn to fly. Thus we arrive at a Socratic acknowledgement of our initial ignorance. In theoretical matters we can only entertain views and defend stand-points, occasionally with intellectual success.

This conception of scientific knowledge does not imply that all foundational problems have essentially been solved. If we accept it, however, we shall at least not be completely taken aback each time an antinomy is discovered.[3]

Scientific philosophy remains in continual and close interaction with the exact sciences, though in recent years it is gaining influence on other sciences as well, in particular on psychology and social science in so far as these disciplines aim at more exact theory formation. Thus scientific philosophy comes to be associated with the exact sciences in particular, whereas the other principal domains of contemporary philosophy are connected with different disciplines. Traditional systematic philosophy may be more or less associated with the older humanities (theology, jurisprudence) and with the exact sciences in their 'classical' form (Euclidean geometry, Newtonian physics), modern irrationalism with biology, psychology and, in recent decades, especially with psychiatry.

I think this already suffices to justify scientific philosophy as a scientific discipline. However, not everybody might be convinced of its philosophical import, especially since its characteristic procedures lend it a '*technical*' and '*specialistic*' aspect which sets it off rather sharply from anything we find in traditional systematic philosophy or in modern irrationalism. However, it would certainly be unphilosophical to judge scientific philosophy by appearances.

Before discussing what I consider decisive arguments in favour of the philosophical import of scientific philosophy, I first want to mention two arguments which personally I do not think very convincing, but which may serve to prepare what follows.

(1) We live in an era dominated by technology and specialisation. Since philosophy ought to reflect the spirit of the age it is only natural that it should now assume a technical and specialistic character.

I consider this a weak argument for two reasons: (i) Never has philosophy, scientific or not, been satisfied with reflecting the spirit of the age; philosophy has always claimed to make a lasting contribution, which implies that it should be judged *sub specie aeternitatis*. (ii) It may be expected of philosophy that it runs counter to the spirit of the age, if necessary, for instance by asserting certain eternal values. Thus the argument would much sooner militate against the philosophical import of scientific philosophy.

(2) The problems dealt with by scientific philosophy have always been included among the topics of philosophy; this alone establishes the philosophical import of scientific philosophy.

Although this observation is quite correct, as an argument it constitutes an *ignoratio elenchi*. Whenever the philosophical import of scientific philosophy is called into question, it is not, as a rule, for the reason that philosophical problems are neglected; the imputation is, rather, that it treats philosophical problems in a *non*-philosophical manner.

I shall now turn to what I consider the main issue. In common with other forms of philosophical thought western philosophy aims at indicating one (or the only) road leading to *wisdom*; they differ in that western philosophy believes the course of this road to be directed, or rather determined, by rationally justified knowledge, i.e., by science. It is not necessary to attempt a definition of wisdom here, since this is not the issue; nor need we go into the question whether wisdom coincides with science in its purest and most exalted form or whether it even transcends the purest and profoundest science. The real issue here is that to the western mind philosophy shows a road leading *through science to wisdom*; this typical way of thinking determines the special character of western philosophy and continues to be the most important stimulant in the pursuit of pure science which is equally characteristic of western civilization.

Now the idea that philosophy could ever become '*scientistic*' is, in my opinion, altogether incompatible with the view just outlined. To accept this idea would be a radical break with a philosophical tradition reaching back to Plato and Aristotle, to Descartes and Leibniz; the ideal of pure science would thereby lose all significance. In this respect scientific philosophy is an organic continuation of philosophical tradition. The unmistakable differences with the older systems do not lie in the conceptions of philosophy and of its ultimate object, namely, wisdom, but rather in the *conception of science*.

The founders of western philosophy had a *contemplative conception of science* (Russell, 1945) and it is this conception which we find again in Pos' plea for 'free contemplation', echoed in Van Melsen's appeal for pre-scientific experience. According to this view (the *locus classicus* is Aristotle's *Posterior Analytics*: see A 2, 72^{a}27; A 3, 72^{b}5; A 7, 75^{a}38; A 28, 87^{a}38), which will be referred to as the *evidence postulate* (cf. p. 76), the sciences are founded on *first principles* that can be intuitively grasped by the human intellect and need not be discovered by means of special scientific research. Accordingly, scientific research can never compel us to change or supplement these principles; it can only increase the extent of our knowledge, never deepen our insight. Therefore the philosopher, the seeker after wisdom, is only concerned with the principles of science.

About a century ago, however, first mathematics, and then other disciplines, were confronted with a *foundational crisis*. It involved a drastic revision of the principles of the disciplines concerned, while at the same time the conviction grew that the necessity of further revision is to be expected as scientific research continues. Views differ as to the consequences of all this for philosophy.

(i) Many students believe that the contemplative conception of science has been invalidated by the foundational crisis. Obviously this is the starting-point of scientific philosophy, which also breaks with tradition in several respects; however, this break seems less radical than the break inevitably facing other trends in philosophy.

(ii) Others think the foundational crisis less serious than appears at first sight. Van Melsen (1954) for instance writes that "the recent development of natural science has revealed the faulty and outdated formulations of former philosophical views". As appears from the views mentioned under (iii) and (iv), however, the radical nature of the foundational crisis has been underestimated.

(iii) 'Positive' or 'intellectual' science, which has arisen from the crisis of foundations, and a 'philosophical discipline' corresponding to the contemplative conception of science, are often simply juxtaposed (cf. p. 30). This 'philosophical discipline' claims to give a *more profound* insight than 'positive' science, though it is as a matter of fact simply a copy of the latter in the stage prior to the foundational crisis. Its very existence implies a mistaken view about the nature of the foundational crisis: this crisis did not arise from the acceptance of a positivistic conception of science but from the internal development in the disciplines concerned. This development has been determined by methodical research suggested by the particular problems of those disciplines.

73

(iv) The terminology employed by Pos, for example when he states that philosophy "aims at something that is absolute and complete" and that its fate is to "keep in contact with the sciences and yet establish its own kind of truth", is sometimes strongly reminiscent of the view described under (iii). Still I believe that his ultimate aim lies elsewhere, namely in the emancipation of philosophy from too close a connection with science. The arguments advanced against the views described under (ii) and (iii) lose their force here; however, I wonder what will be the practical consequences of this conception of philosophy.

Like everybody else, the philosopher is in many respects dependent on *information*, so that the question arises where he finds it. The sole sources that I can think of are: (a) pre-scientific experience, (b) science, and (c) possibly a philosophical intuition. Now pre-scientific experience is essentially incomplete, dependent and accidental to such an extent that any philosopher using it as his starting-point will sooner or later be compelled to draw on science as well. Then he will have to reflect on the foundations on which science rests. A philosophical intuition, if there is such a thing, is either '*private*' or '*public*'. Every available 'public' source of knowledge, however, is also exploited by science, whereas an appeal to a 'private' source of knowledge implies a radical break with western tradition. Thus if philosophers want to avoid such a break they are obliged to maintain a close contact with science.

(v) Finally I wish to mention an attitude which is rare nowadays but was very common in a not so distant past. This attitude involves the rejection of all scientific results that are connected with the crisis of foundations.

Taking all this into consideration, I think the position of scientific philosophy is firm enough. However, the question whether or not it is in a position to show us the road to *wisdom* is still unanswered. Even its basic principles cannot claim the unassailability, the absoluteness and the completeness characteristic of, or at least aimed at by, traditional philosophy.

I should like to say the following about this objection. In so far as the pursuit of traditional philosophy leads to wisdom, this cannot mean that it furnishes infallible knowledge of first principles. The feeling of having penetrated to the deepest roots of things may, to be sure, give strong satisfaction, but this intellectual sensation of power must not be equated with wisdom. It would seem to me that wisdom is rather the fruit of the spiritual concentration necessary to *discover* first principles, and more in particular, to distinguish true from false principles. In this respect, however, the student of scientific philosophy has to struggle with exactly the same problems; he will have to take a stand on a variety of problems closely related to questions of traditional philosophy. I think that the possibility of attaining wisdom along this road cannot reasonably be denied.

Of course this does not mean that the student of scientific philosophy would be able to show the one and only road leading to wisdom, or *a* road that will lead us there infallibly and without effort. Spinoza's word still holds true: whatever is excellent is at the same time difficult and rare.

REFERENCES

[1] This chapter comprises the introduction and final section of the original text. I have omitted the middle part, which contained a number of examples, and the Postscript.

[2] F. Gonseth, 'La preuve dans les sciences du réel', in *Théorie de la preuve*, 1954.

[3] P. Bernays, 'Zur Beurteilung der Situation in der beweistheoretischen Forschung', in *Théorie de la preuve*, 1954.

IX

MODERNISM IN SCIENCE

1. THE EVIDENCE POSTULATE

We all know the passage from Plato's *Meno* where Socrates, just by asking all kinds of questions, succeeds in making a young slave discover for himself a simple geometrical theorem.

I do not want to discuss the background and implications of this passage here. It may, however, serve to characterize an idea according to which geometry, as indeed any scientific discipline, should start from *fundamental concepts* and *propositions* the meaning of which is clear without further definition, the truth of which is evident without further demonstration. Thus these *principles* do not have to be discovered by means of deliberate and systematic research. Every human being has command of them; their mastery is no monopoly of the scientifically trained man or the philosopher.

The extremely plausible view of the foundations of science that finds expression in this so-called *evidence postulate* lies at the root of Aristotle's philosophy of science, which until 1800 was universally accepted. This philosophy of science is implicit in the various systems of philosophy that have been developed since Plato and Aristotle. It has, more specifically, conditioned the kind of problems typical of metaphysics and epistemology. Subtleties and details apart, the various schools of philosophy hardly differ in their ways of stating the problems; disagreement only begins when solutions are proposed.

Aristotle's philosophy of science is attractive in guaranteeing the *unity of science*; since it also applies to disciplines like ethics and aesthetics, it ensures at the same time the *unity of culture*. Indeed, if all scientific disciplines (and thus, indirectly, morality and law, technology and the fine arts as well) are founded on principles that are universally valid and unchanging, principles with which every human being is (or could make himself) familiar, then the various domains of culture will, as long as the right principles are borne in mind, necessarily develop along harmonious lines.

2. THE PLURALISM OF MODERN CIVILIZATION

A few years ago, an art scholar asked the head of a mathematics department for the loan of some mathematical models. He wanted to use them
to enliven an exposition of contemporary abstract art. In principle the
mathematician was willing to co-operate. However, he had to make a
reservation with regard to the condition the available material was in.
The fact was that the models had hardly been used ever since teaching
had gradually been adapted to the development of modern abstract
mathematics.

The state of affairs characterized by this anecdote would seem to be
the result of the development of mathematics, and of numerous other
domains of culture, since 1800. The moral is, first, that contemporary
mathematics is based on principles which even for a modernistically
minded student of a different scientific discipline, who is moreover an
interested party, are not immediately accessible. Further, it appears that
it has become impossible to indicate a link between the different domains
of contemporary civilization. Both mathematics and art have gone
through a radical development, but there is no reason to believe that the
two developments were in harmony with each other.

Of course nothing is proved by an anecdote, and we may take comfort
in realizing that in the past, too, adequate knowledge of even the elements
of a discipline like mathematics, though perhaps attainable for everybody
in principle, in practice was difficult to acquire for outsiders. Further,
it may be pleaded that already in the past developments in different
domains of civilization have by no means always been in tune, or even
more or less parallel.

3. MODERNISM IN SCIENCE

These reflections, however, are not conclusive. This becomes apparent
when we consider a little more closely what I should like to call *modernism
in science*. Since 1800 there certainly has been a development in the
domain of science which forces us to reconsider traditional philosophical
views and ways of stating problems, and thus to make a fresh start in
thinking about our civilization in its entirety.

However, the purport of my argument inevitably involves the following difficulty. If it is true, as I hope to show, that modern science is

based on principles that *cannot* be considered to be common property of mankind, as the evidence postulate of Plato and Aristotle would have it, then these principles will not lend themselves to a plain exposition. On the other hand, my argument can hardly do without a comparison of the modern principles with the traditional ones. By taking the matter of my argument from mathematics, I am going to make my task even harder than would seem strictly necessary. As a matter of fact, however, this subject serves my purpose best, since here more than anywhere else the principles have evolved in a most spectacular way.

4. EUCLID'S GEOMETRY

The geometry of Euclid, which in the main corresponds with mathematics as it is still taught in schools, may serve as a representative of traditional mathematics. This mathematical discipline may be characterized as *abstract*, *deductive*, and *intuitive*. I use the term 'abstract' here in conformity with Aristotelian tradition, i.e., in its proper sense. For the figures and solids studied in schoolbook mathematics may be thought of as having been obtained from real, material and observable objects, by *abstraction* from qualities like colour, weight and temperature.

On account of this abstraction the figures and solids of geometry are not open to empirical examination, so that only the *deductive method* remains. On the other hand, having been obtained from observable objects, geometrical figures and solids are intuitive to a high degree.

Thus Euclid's geometry starts from fundamental concepts and fundamental propositions (or axioms) that are founded on intuition; logical conclusions are drawn from these principles by means of deduction. In other words, the evidence of geometrical propositions derives partly from intuition, partly from logic.

This dualistic character of geometrical evidence eventually failed to satisfy students of pure mathematics. The discovery, round about 1830, of non-Euclidean geometry no doubt contributed greatly towards the general feeling of uneasiness that arose; nor did the difficulties which mathematicians experienced in trying to find a foundation for the infinitesimal calculus fail to have some influence. After 1848 especially, ever more resolute attempts were made to eliminate the intuitive foundation, and thus at the same time the appeal to Aristotelian abstraction, and to arrive at a purely logico-deductive construction – not only for geometry, but for the whole of pure mathematics.

A construction of this kind is meant when we talk about modern abstract mathematics. Thus the term 'abstract' is not used in its strict sense in this connection.

5. SET THEORY

In its most radical form abstract mathematics uses the concept of *set* as the sole fundamental concept.[1] This might seem absurd, since according to common usage it is impossible to talk about sets unless we also have at our disposal objects that may be taken together to form sets or, in mathematical terms, that can occur as *elements* of sets.

In this customary sense the concept of set indeed occurs already in elementary geometry, albeit under a different name: what goes by the name of *loci* in elementary mathematics are called *point sets* by mathematicians.

The objection just mentioned is met by abstract mathematics in a way that will seem rather curious at first sight: sets are themselves allowed to occur as elements of sets. Frequently even a further step is taken: objects that are not sets are just left out of account. That this *set-theoretic monism*[2] actually leads somewhere will become clear from the following observations, which will also give an impression of mathematical reasoning in set theory.

(a) Let 0 be the set which has as elements all sets not occurring as an element of any set. Then the set 0 (the *empty* or *null* set) does not contain any element.

In proof of this, let x be any set and suppose x is an element of 0. In order to belong to 0, x must be a set not occurring as an element of any set; in particular x must not be an element of 0. Since this contradicts our initial hypothesis, this hypothesis must be rejected. Hence the set 0 cannot contain any element.

(b) Let 1, or $\{0\}$, be the set containing the set 0 as its sole element. Then 1 differs from 0, since 0 has been proved to contain no element, whereas 1 does contain an element, viz., 0.

(c) Let 2, or $\{0, 1\}$, be the set containing the sets 0 and 1 as its only elements. Then 2 differs from 0 as well as from 1, for 0 contains no element, whereas 2 contains, e.g., the element 0: this establishes that 2 differs from 0; further, 1 is not an element of 1, whereas it is an element of 2: thus 2 is seen to differ from 1.

Obviously we may now go on to construct the sets 3, 4, ..., differing from the sets 0, 1, 2 as well as among themselves.

(d) Let Z_0 or $\{0, 1, 2, 3, 4, ...\}$ be the set having as elements all the sets 0, 1, 2, 3, 4, ... Observe that Z_0 contains infinitely many elements.

The formation of the sets 0, 1, 2, 3, 4, ..., Z_0 (and of many other sets) can be based on the following principle, which is known as the *comprehension axiom* (see, e.g., Beth, 1959, pp. 465 ff.):

Well-defined objects sharing a certain characteristic property constitute a *set* of which they are the *elements* and which is uniquely determined by the property concerned. Every set is a well-defined object.

Disregarding various additions, corrections and refinements that would be required in a more systematic exposition, we may now characterize *pure* or *abstract set theory* as a deductive theory which has the concept of *set* for its sole fundamental concept and the *comprehension axiom* for its sole fundamental principle.

Set theory, which originated from the ideas and the work of Bolzano (1781–1848), Cantor (1845–1918), and Zermelo (1871–1953), differs greatly in character from Euclidean geometry. It rests upon principles with which certainly not every human being is acquainted and thus does not answer to Aristotle's conception of science.

It might indeed be supposed that the fundamental concept of set has been obtained by abstraction from some generally used concept of set. This view proves to be untenable. If we apply the comprehension axiom to concrete objects, we can immediately form the set of all genuine paintings by Rembrandt or the set of all postage stamps issued since 1840. These cases do not, however, constitute sets in the sense of common usage.

Accordingly the comprehension axiom is to be considered far from self-evident. The introduction of the natural numbers 0, 1, 2, 3, 4, ... strongly suggests a mathematical swindle. Suppose I am penniless, but I find a little note-book, a pencil and an empty tobacco-tin on the pavement. Now I first make up the cash and then draw up a balance-sheet, mentioning an office inventory and goodwill as the assets. These assets are then 'subsumed' under a limited liability company.

The construction of abstract set theory strongly reminds one of this way of doing 'business', and did not in fact inspire much confidence at first. This is all the more remarkable as in Cantor's or even in Zermelo's work set theory had not by far assumed the radical form characterizing it today; set-theoretic monism in particular is not yet present in their work. Although even now opposition has not been completely silenced, the significance of set theory – in its most radical form – is receiving more and more acknowledgement.

Obviously this acknowledgement cannot be attributed to the self-evidence of the first principles. It rather springs from the recognition that set theory provides a suitable basis for a coherent systematization of the strongly divergent deductive disciplines which together constitute modern mathematics. Apart from this, however, we are impressed by the fascinating beauty of abstract set theory and the unparalleled audacity of the constructions which it allows.

6. PHILOSOPHICAL CONSEQUENCES

Traditional epistemology, in accordance with Aristotle's conception of science, had never questioned that pure mathematics, in its pre-1800 stage, is based on clear and evident principles with which every human being is acquainted. Its objective was to elucidate the origins of this knowledge and to explain the self-evidence of the principles. Considering the attention that was given to mathematics by the great thinkers of the past, we cannot rid ourselves of the impression that the questions pertaining to the foundations of pure mathematics must be of great importance to traditional systematic philosophy.

If this impression is correct, then the development of pure mathematics since 1800 – and in particular of abstract set theory which, as we saw, does not satisfy Plato's and Aristotle's evidence postulate – must be a severe blow to traditional systematic philosophy. Indeed the danger was realized in philosophical circles, as appears from the fact that in the course of the 19th century the developments in pure mathematics have time and again met with vehement polemics by leading figures in systematic philosophy. By way of example I quote a pronouncement on non-

Euclidean geometry, sometimes called *absolute geometry*, by the Dutch Hegelian G. J. P. J. Bolland:

> However, absolute mathematics is of little avail and has thus far fostered only the vagaries of fancy. ... Our native abilities are such that we can never meet with anything but 'Euclid-objective' simultaneities, and an 'absolute mathematics' which claims to be more than an innocent and rather unprofitable intellectual game is an illusion of the worst kind. ... 'Mathematical' effusions like these do not tend to inspire much confidence in the judgment of mathematicians (Bolland, 1889).

This polemic did not produce much effect. Systematic philosophy has broken up into various schools which each view the contemporary situation from their own stand-point. I shall mention only a few of these schools.

(i) A reactionary group simply refuses to recognize the new situation or considers it philosophically irrelevant. This attitude enables representatives of various schools to cling to one of the traditional systems of philosophy.

(ii) Some phenomenologists and many analytical philosophers are attempting to conceive a new, less ambitious theory of knowledge which is concerned with the contents of the natural consciousness and with the expression of these contents in terms of ordinary language; it is not concerned with scientific thinking and reasoning.

(iii) Many representatives of the existentialist camp continue to employ the traditional phraseology of systematic philosophy, but have given up all claims of providing a systematic foundation for the unity of science and thus for the unity of civilization.

(iv) Finally there is an attempt to rebuild systematic philosophy in close association with science in its modern form. In a radical form this tendency is especially manifest in *neopositivism*, as promulgated in particular by the 'Wiener Kreis'. *Logicism*, which goes back to Frege and at present is advocated, e.g., by Church (in a Platonistic sense) and by Carnap and Quine (in a nominalistic sense), is somewhat closer to traditional views (cf. Beth, 1959).

It need hardly be argued that under these circumstances traditional systematic philosophy has lost a considerable amount of its former prestige.

7. CONSEQUENCES FOR CIVILIZATION IN GENERAL

It is even more important, however, to realize that contemporary philosophy cannot be expected to be an effective guarantee for the unity of civilization. The schools mentioned under (i)–(iii) quite consciously sever the traditional links between philosophy and science. On the other hand neopositivism, which does aim at furthering the unity of science, can hardly be expected to supply a normative ethics or aesthetics. The fact that several representatives of the schools here referred to have not only contributed to philosophy but also to art and literature is in itself quite gratifying, but does not ensure the unity of civilization.

Thus the question arises whether this might not be achieved by other unifying forces in our civilization. A detailed investigation into this question is of course not possible in the compass of this essay. I want to restrict myself to a summary discussion of one particular aspect which has already been touched upon earlier.

The more or less simultaneous emergence of modern abstract mathematics and contemporary abstract art, and the mathematical associations of the latter, would of course suggest a certain congeniality and also some factual link between the two phenomena. In conclusion to this essay therefore, I should like to discuss the following questions:

(i) Is there indeed some degree of congeniality or at least a certain analogy between the phenomena just mentioned?

(ii) If some connection or analogy can be found, does this indicate a common origin, a mutual influence or some other factual link?

(iii) If a factual link can be shown to exist, is this perhaps an indication of the influence of unifying forces which might at the same time ensure the unity of civilization?

As concerns question (i): I think it would be very difficult to point out some congeniality between abstract mathematics and abstract art, though we may easily find an analogy. Both have resulted from a conscious and radical break with traditional principles, standards and means of expression, and for that reason both are rather inaccessible to those who have been 'brought up with' tradition.

Having pointed out this analogy I am obviously obliged to take up question (ii). By a factual link we may understand in this connection: (a) a common origin, (b) a mutual influence, (c) an influence of art upon mathematics; or (d) an influence of mathematics upon art. Now it should be kept in mind that abstract mathematics has developed very gradually since 1847, and that this development was called forth by problems reaching back much further into the past, whereas abstract art is one among several aspects of artistic modernism after 1900. Thus the possibilities mentioned under (a), (b) and (c) need not be taken into account.

The possibility mentioned under (d) seems more promising, especially since mathematics is often referred to in theoretical discussions about abstract art. If we go into the matter a little further, however, we appear once more to have turned into a blind alley. The *plastic mathematics* of M. H. J. Schoenmaekers (Schoenmaekers, 1916) which figures so prominently in the theories of *De Stijl* (Jaffé, 1956) is no genuine mathematics at all.

We do find genuine mathematics in Le Corbusier's sketches and notes for the Philips Pavilion at the Brussels World Exposition (Le Corbusier, n.d.). However, these observations entirely belong to what is usually called *classical mathematics*, as opposed to abstract mathematics; in other words, they belong to the mathematics of plaster and wire models.[3]

Indeed it is not surprising that a contemporary artist, no matter how abstract his inclinations may be, feels attracted to classical rather than abstract mathematics. The mathematical disciplines with which he is brought into contact during his training because they are technically relevant for him, belong entirely to classical mathematics; as a result, moreover, of his special talents and of the use he has to make of these disciplines, their intuitive aspect will stand out most prominently for him. Conversely, abstract mathematics will not by comparison be easily accessible for him and cannot, therefore, be expected to influence his ideas and his creative work.

Hence the question mentioned under (ii) will have to be answered in the negative, and we may accordingly refrain from discussing question (iii). Furthermore, reverting to question (i), we now realize that the analogy that was observed is wholly superficial and that a fundamental relationship between abstract mathematics and abstract art would seem out of the question.

Our discussion, then, has not brought to light any unifying forces but rather emphasized the pluralistic nature of modern civilization. For those who are under the spell of cultural fatalism (cf. pp. 1ff.) this result will no doubt be quite alarming. According to Frobenius, Lamprecht, Pareto, and Spengler, every viable civilization is an organic unit; thus the pluralistic nature of our civilization would have to be interpreted as a symptom of its approaching downfall. This pessimism appears the more justified since philosophy has forsaken its traditional normative task.

I think there is very little cause, however, to surrender to such prophecies.[4] Discrepancies of development in different areas within one and the same civilization are such a common phenomenon[5] that it simply won't do to attach too much importance to it, especially in a case like the one discussed above, where it appeared that the discrepancy could be so easily explained. I would even take one step further and express the opinion that the pluralistic nature of modern civilization should not be deplored at all but should on the contrary be highly valued.

REFERENCES

[1] Matters have been simplified here. It is the relation between an object and the set containing this object as an element which is the fundamental concept.

[2] As an introduction to set theory Halmos (1960) is to be recommended. An extensive discussion of its foundations is to be found in Fraenkel and Bar-Hillel (1958).

[3] In modern mathematics the term '*model*' usually denotes constructions in abstract set theory.

[4] Cp. Popper, 'Prediction and Prophecy and their Significance for Social Theory', in *Proceedings* (1949).

[5] The opposite is often maintained without adequate factual support, as is demonstrated by Dijksterhuis (1956).

X

MATHEMATICS AND MODERN ART

It may occur that a philosopher, in setting foot on territory with which he is not yet familiar, imagines to be following a sudden, almost or entirely unaccountable impulse and that only after some time he comes to realize how compellingly the choice of a fresh object for his reflections was forced upon him by a tendency in his philosophical pursuits that has been prevalent for a long time. The preceding essay on 'Modernism in Science', which was written about a year ago, now seems to me to provide a characteristic example of this phenomenon.

My philosophical pursuits are aimed at helping to create the conditions for a rational discussion of the most diverse spheres of human activity and experience. So far I have in this respect been mainly concerned with logic and the foundations of various domains of science. The rationale of this restriction will hardly need explaining. Logic is a precious instrument in any rational discussion, and the foundation of science lends itself to rational discussion more readily than any other subject.

However, I do not consider this restriction definitive. It is my firm conviction that other domains of human activity and experience, too, must and can be made accessible for rational discussion (cf. Beth, 1957b).

A statement like this usually provokes diffidence and even protest. Therefore I at least want to make an attempt to preclude at the very outset the misconceptions that are almost unavoidable in this context. In my opinion, the aim of rational discussion on artistic matters, for example, could never be to give precepts for the manufacture of works of art or to lay down standards by which to judge them, although unfortunately this has often been assumed. Rational discussion has no other than the quite modest purpose to promote the possibility of rational communication in questions concerning, for instance, the manufacture and the valuation of works of art.

Now of course the question may be raised why I think it desirable for this communication to be rational, i.e., to be such that an appeal is made to the common sense of our partner in the discussion, to his willingness to take into consideration any views that are brought forward and to enter into the arguments adduced in support of these views. My reply to this question is simply this: I prefer rational communication to all other kinds of mutual or one-sided influencing, as it is the only kind of influencing that does not violate the freedom of our fellow-men.

Of course I am well aware that we cannot confine ourselves to an appeal to reason under all circumstances. However, the appeal to 'guidance', to authority, to compulsion, to violence is always a concession to the *de facto* unreasonableness of man and thus is very much out of place where intellectual or spiritual or other lofty matters are concerned.

The argument is not an unfamiliar one, and I am afraid it can hardly be repeated without giving rise to a variety of objections that could with so me justification be made against older forms of rationalism. But I shall resist the temptation to safeguard myself against such objections and come to my point. My claim that rational discussion is not out of place even in art surely presents a shrill contrast to the nature of the discussions that nowadays are prevalent in this field.

Of course not everything must be judged by one and the same standard in this respect. The creative artist trying to make us sympathize with the artistic aspirations of the group of which he is the spokesman, is completely justified if he expresses himself in an aphoristic, emotional and suggestive way. The critic, writing with the pretension to inform the public about these aspirations, has no right to express himself in this manner. Unfortunately however, it is all too common for critics to appeal to some unverifiable knowledge which they possess or to irrelevant emotions of his public.

In this connection, as becomes clear to me now, the theme of my essay 'Modernism in Science' had the considerable advantage of inviting rational discussion. It can hardly be doubted that the problem of the relations between modern abstract mathematics and modern abstract art is completely unsuitable for any but a rational treatment.

In order not to repeat myself I should like to approach the subject of 'Mathematics and Modern Art' from a different angle than in the preceding chapter. It would seem, then, that the real, possible or putative relations between mathematics and art may be subsumed under three categories, namely

(1) the application of mathematics to art;

(2) ideas[1] which are naturally common to mathematics and art;

(3) the adoption of mathematical ideas in art.

In the cases referred to under (1) we are dealing with applied mathematics as it occurs, on a much larger scale, in natural science and technology. The most typical examples are to be found in architecture, as a result of its highly 'technical' character. These cases are not necessarily interesting for our purpose, since the application of a fragment of mathematics need not imply that the artist is aware of the underlying ideas. Yet they may appear to be relevant in so far as they may call the artist's attention to ideas as referred to under (2) and (3). Indeed I suspect that his interest in such ideas is generally roused in this way; this is at least suggested by the examples to be given presently.

Category (2) comprises ideas like *symmetry* and *perspective* (M. Kline, 'Projective Geometry', and H. Weyl, 'Symmetry', both in Newman, 1956). As for the idea of symmetry, this will not need to be further illustrated, but I should like to say a few words about perspective. Even if true perspective is not a postulate of general aesthetics, it may be a requirement of a particular artistic style. In order to make this requirement, or in order to fulfil it with a degree of approximation that is aesthetically satisfactory, we do not need a mathematical theory of perspective. Of course the artist will consult the theory of perspective, if such a theory has been developed[2], but then we are again in a situation as referred to under (1). Conversely, the mathematician will not need any exhortation on the part of art in order to arrive at the development of a theory of perspective.

Unfortunately I have never come across a good example of what was meant by category (3). I restrict myself to mentioning, albeit hesitatingly and with reservations, the *golden section* (Panofsky, 1955).

Obviously other possibilities than those mentioned under (1)–(3) may be imagined. In particular I mention the possibilities of the adoption of artistic ideas in mathematics and of the adoption, in both mathematics and art, of ideas from some third domain. However, I am unable to quote examples of this, and we shall see, moreover, that these possibilities, which were mentioned for the sake of completeness, have little bearing on the concrete problem with which we want to deal here.

The fact that the developments of mathematics into abstract mathematics and of visual art into abstract art have been practically simultaneous, together with the often mathematical (or quasi-mathematical) character of abstract art, suggests a certain connection. We need not consider here any possibilities other than those referred to under (1)–(3). For modern abstract mathematics came into existence between 1830 and 1850 and is thus older than abstract art; accordingly abstract art, as we already observed, often has a strikingly mathematical character, whereas abstract mathematics shows no traces of having been influenced by visual art. And finally, theoreticians of modern abstract art have repeatedly referred to mathematics.[3]

Firstly, then, I wish to consider the question whether the development of abstract art is partly or wholly due to the adoption of ideas from abstract mathematics. Then I shall discuss the possibility that mathematics and art have in common something that might be called the *idea of abstraction*, and that as a result of certain circumstances this idea has manifested itself simultaneously in the two fields.

Since the first question was considered in some detail in my essay on 'Modernism in Science', I shall only deal with it summarily here. The theoretical foundations of abstract art could be partly identified comparatively easily with the help of H. L. C. Jaffé's well-known study on *De Stijl* (Jaffé, 1956). The references to mathematics are either to what is now called *classical mathematics* or to M. H. J. Schoenmaekers' *"plastic mathematics"* (Schoenmaekers, 1916) – which is not authentic mathematics at all – but *never* to modern abstract mathematics.[4]

These views were not confined to the *Stijl*-group. In W. Kandinsky's *Punkt und Linie zu Fläche* (1926) there are many references to mathematics. Most of the numerous figures, however, are of little significance from a mathematical point of view, and they all belong to classical mathematics. In an earlier work by the same author, *Über das Geistige in der Kunst* (1912), science turns out to be represented by people like Blavatsky, Flammarion and Steiner. In his work *Über die moderne Kunst* (1945) Paul Klee confines himself to one or two allusions to more-dimensional geometry and to Minkowski's theory of the space-time continuum. Thus even in these representative authors we fail to discover any sign of a direct impact of abstract mathematics on abstract art.

At this point I should perhaps try to stave off a misunderstanding that might easily arise. It is not my intention to reproach the authors and artists mentioned above with not having made a study of modern abstract mathematics, or with having failed to be inspired by modern mathematics in their creative work and in their writings. I simply want to point out that there is no factual basis for the frequently expressed supposition of an impact of abstract mathematics on abstract art.

The discussion of a possible *idea of abstraction* will lead us to a somewhat more positive and thus more satisfactory result. I wish to make it clear from the outset that I shall use the term *'abstract'* naively, irrespective of any epistemological foundation. In this sense we may for instance say that the figures of classical geometry are *'abstract'* as compared with the objects in the world of experience to which they manifestly correspond. In the same naïve sense we may also observe that the 'sets' dealt with by modern abstract mathematics are *'abstract'* in comparison with the figures of classical geometry. All this would support the view that the *idea of abstraction* is inherent in mathematical thought and that this idea manifests itself in mathematical objects becoming ever more *'abstract'*.

With respect to art I refer to Worringer (Worringer, 1908), who in "*das Kunstwollen*" (the artistic impulse) identifies the "*Abstraktionsdrang*" (urge towards abstraction) as the antipole of the "*Einfühlungsdrang*" (urge towards empathy). With the emergence of science, abstraction lost its hold on art and it became possible for classical art to evolve. "Its aspiration is no longer the rigid lawfulness of the abstract, but the gently harmony of organic being."

The fact that the classical forms of art are superseded by abstract art, as appears to be characteristic for our age, does not quite tally with Worringer's view in its original form. However, on account of the evidence produced by him we seem to be quite justified in assuming that the *idea of abstraction*, like the idea of perspective, is inherent in visual art as well as in mathematics. Thus we are not, in my opinion, compelled to rule out the possibility that this idea, after having been latent for centuries, once more becomes manifest at a certain moment.

Even though we accept this possibility, however, we must not expect to find a deeper cause of the parallelism between the developments in mathematics and visual art. Indeed if we survey a longer period of time, we see that this parallelism does not continue. In mathematics we find a continual increase in abstraction, whereas visual art constantly oscillates between the two poles of 'empathy' and 'abstraction', in Worringer's terminology.

Yet this idea of parallelism is cherished by many people. In the first place, quite naturally, by those who want to conceive all that occurs in the various domains of human civilization during a certain period as manifestations of the 'spirit of the age' ('*Zeitgeist*'; cf. Dijksterhuis, 1956); and in the second place, by the apologists of contemporary art.

As for the theory of the spirit of the age, I just want to observe that this theory will appear to hold good as long as simultaneous developments in related subjects are compared, whereas a comparison of subjects that are less closely related often yields a completely negative result. An example is offered by our discussion of mathematics and modern art.

The apologetics of contemporary art is a most curious phenomenon, certainly deserving a more thorough investigation than is made here. Usually the theory of the spirit of the age is chosen as a starting-point. Then one of two kinds of arguments is used. According to the first kind (Kandinsky, 1912, p. 11), creative minds are as a rule ahead of their time and therefore they are generally misunderstood by their less creative contemporaries. From the fact that there appears to be so little appreciation for contemporary art it is then concluded that this art must have emanated from creative minds. Obviously the characteristic nature of contemporary art does not play an essential part in this kind of argument.

Matters are different according to the second kind of argument. Here the abstract nature of contemporary art is to explain why it is often so difficult of access. The term 'abstract' is taken in a very enlarged sense here, but this does not seem to me to be essential; even if one particular school, not wholly without justification, claims the exclusive right to the label '*abstract art*', this does not alter the fact that other artistic manifestations of our time also have an '*abstract*' character, and that this is why they are less easily accessible. With an appeal to the spirit of the age the argument then reminds us of the abstract character of contemporary science, and thus the abstract character of contemporary art is thought to have been accounted for.

Obviously both arguments are formally highly questionable, but there are far more serious objections. One of the most deplorable aspects is the implied scorn for the freedom of (the individuals constituting) 'the public'. They are hit in a weak spot, namely their uncertainty as to how contemporary artistic manifestations are to be judged. According to the first argument this uncertainty indicates a lack of creativity; the second argument attributes the uncertainty to an incapacity of abstract thought. Thus both arguments play upon latent feelings of inferiority. In the long run this kind of apologetics cannot but positively injure the already precarious relations between art and the general public.

Little is to be expected of arguing against the thus stigmatized apologetics. It will be more rewarding to try to gain some insight into the ulterior motives, not only of the apologetics of contemporary art, but also of those reactions of 'the public' at which this apologetics is aimed, and perhaps even (though this is said with many reservations) into the sources from which spring the less approachable artistic manifestations which evoke these reactions.

In making such an attempt I should like to start from the hypothesis that these ulterior motives are to be found in *feelings of uncertainty with regard to aesthetic norms and values*, which are very wide-spread nowadays (Kandinsky, 1912, p. 18).

There are two considerations supporting this hypothesis:

(1) the prevalence of feelings of uncertainty can be explained by our knowledge of a variety of circumstances;

(2) it is also quite plausible that these feelings should lead, not only to reactions on the part of 'the public' like those described above, but also to the apologetics of contemporary art.

Re (1). Every human being has an innate sense of beauty. Since this statement may give rise to misunderstanding, some further explication is required. I certainly do not mean to say that we are born with an innate set of aesthetic norms and values. I do however assume that man is by nature subject to spontaneous reactions and types of behaviour in which a certain aesthetic preference can manifest itself.

The aesthetic norms and values which are more or less universally accepted in a certain period are due to an interplay of our innate sense of beauty, our social and natural surroundings and the existing traditions of society. This interplay can only then lead to a stable result if the conditions of life do not change too rapidly. On the basis of his innate sense of beauty, each individual then has the opportunity to acquire a sense of aesthetic norms and values, which may be manifested as good taste, artistic judgment or artistic creativity.

The feelings of uncertainty noted above suggest that an unwavering sense of aesthetic norms and values is lacking, that the interplay is disturbed. It is, in my view, quite understandable that such a disturbance should occur in our time. Circumstances are changing in an increasingly rapid pace. As a result of the pluralistic nature of our civilization[5], the individual is confronted with a variety of divergent traditions. Moreover, the amount of ugliness which modern industrial civilization, especially in the early stages of its development, has inflicted on the human race, may have atrophied the innate sense of beauty of several generations.

Re (2). Even in default of an unwavering sense of aesthetic norms and values, we are still, as a rule, not unable to form a judgment when we are confronted with the art of a not too distant past, since its interpretation and its value are determined by tradition. In the case, however, of novel and moreover rather inaccessible works of art, tradition soon fails us (see Kandinsky, 1912, p. 103). For most people the result will be that their latent feelings of uncertainty become manifest and that they become amenable to apologetics.

There are some, however, with a natural disposition that renders them likely to react in a different manner. As a result of their confrontation with certain forms of art they may develop a new sense of aesthetic norms and values. This sense of values does not usually develop organically: it continues to be heavily burdened by feelings of uncertainty and thus to be slightly constrained. This may give rise to an apologetics which endeavours to silence the inner uncertainty by 'converting' other people.

Even if the above attempt to sketch the way in which the required insight may be obtained has been successful, I should like to emphasize that the existing uncertainty and the apologetics engendered by it have by no means been eliminated yet. However, the foundations have been laid for a rational discussion of the aesthetic problems of our time and thus for rational communication, which will be able to replace much unprofitable apologetics. Rational communication might in the long run help to create the conditions of life which man needs in order to acquire a renewed and deepened sense of aesthetic norms and values.

REFERENCES

[1] The term 'idea' has been used in this essay to indicate a notion which can of course be conceptually defined only within each separate discipline. Still it is of paramount importance in quite divergent fields and, moreover, remains identifiable as 'the same'

notion when we move from one discipline to another.

2 See E. Panofsky, 'Dürer as a Mathematician', in Newman (1956).

3 Cf. Jaffé (1956), Schoenmakers (1916), Kandinsky (1912, 1926), and Klee (1945).

4 In Essay IX a sketch has been given of modern abstract mathematics. Cf. Van der Waerden (1928).

5 Cf. pp. 77 and 84–5.

XI

IN RETROSPECT

From time to time we stop to cast a glance at the road lying behind us. It is not unbecoming for a philosopher to indulge in this human propensity, although he will have to reflect whether he is justified in bothering other people with the outcome of such retrospective ponderings, since even if they are not merely anecdotal, they are more often of a subjectively autobiographical than of an objectively historical nature.

It would seem to me that the publication of the ensuing reflections may perhaps be justified by the circumstance that the personal experiences to which they refer may be considered as more or less typical. It is quite probable that others have had similar or at least comparable experiences, so that an account of my own development may contribute towards a clearer insight into the origins of what is sometimes called the intellectual climate of a period.

I still retain direct, although quickly fading, memories of the years between about 1924 and 1934, a period immediately preceding my first independent philosophical activities. Nobody, I think, will be wronged if we consider this period as a time of transition and of – real or apparent – calm. In Holland the aspiration to find a religious or at least ideological foundation for the development of philosophy formed the centre of interest. It goes without saying that such concentration upon essentially non-philosophical problems involved the risk of a narrowing of the horizon and of one-sidedness.

One of the great merits of the 'Genootschap voor Critische Philosophie' (Association for Critical Philosophy), founded at Utrecht in 1923, was to have pointed out the distinction between philosophy and ideology (in the broad sense of 'Weltanschauung'), and to have defended the autonomy of philosophical thought. During more than 10 years its leading figure, Goedewaagen, succeeded in keeping together in close collaboration a number of like-minded, mostly young philosophers. Even though his conception of the autonomy of philosophical thought was very dogmatic, in accordance with Marburg neo-Kantianism, I do not think that this detracts from his other merits. This dogmatism prevailed only *de facto*, not *de jure*. It was always permissible to doubt or contradict. Thus in my very first publications, shortly after I became a member of the 'Genootschap' in 1933, I already warned against "an overly dogmatic conception of the fundamental principle of autonomy". Apart from some lectures given by B. J. H. Ovink and A. J. de Sopper, I owe my philosophical training largely to my membership of the 'Genootschap'. Later on, the exclusively Marburgian leanings were abandoned and in 1938 the name 'Genootschap voor Wetenschappelijke Philosophie' (Association for Scientific Philosophy) was adopted. The immediate cause was the crisis brought about by Goedewaagen's resignation in 1937, but ultimately the development of Dutch philosophy, which was partly due to the 'Genootschap' itself and may be summarily described as a turn from *Weltanschauung* to philosophy, was responsible for the change.

Indeed the very fact that the 'Genootschap' initially wished to defend the autonomy of philosophy also with respect to the various branches of science led it to consider the quite opposite tendency, i.e., the tendency to establish a very close association between philosophy and the sciences. This tendency, which had originated mainly in connection with the investigations into the foundations of the exact sciences, was especially prevalent, round about 1930, among the members of the 'Wiener Kreis'. Even though their points of view were not always accepted without qualification, the results of the foundational research as well as the work of the 'Wiener Kreis' were studied very thoroughly and discussed extensively. This aspect of the work of the 'Genootschap' became clearly manifest during the Winter Conference held at Utrecht on January 15, 1933, where papers were read by P. H. van der Gulden on 'Symbolic Logic' and by P. G. J. Vredenduin on 'Philosophy of Mathematics'. Thus the 'Genootschap' became a medium through which Dutch philosophers could once more establish communication, not just with like-minded groups in other countries, but with the international philosophical world at large. Other media were the signific movement, the journal *Synthese* and, at an international level, the philosophical congresses at Prague (1933) and Paris (1937).

Although Dutch philosophy fortunately retained its distinctive features in many respects, it soon lost its somewhat provincial character as a result of these developments. Partly owing to the activities of Leo Polak and H. J. Pos, this was widely recognized, as appears from the fact that during the Paris congress the Netherlands were charged with the organization of an international philosophical congress in Groningen in 1941. This plan could not be carried out till 1948, when the Tenth International Congress of Philosophy was organized in Amsterdam. The renewed interest in the scientific element also benefited the exchange of ideas in this country and once more, now in a larger circle, focussed the attention on the achievements of Dutch scholars. Only then did Brouwer's intuitionism, Mannoury's significs and Heymans' ideas receive the more general regard to which they were entitled.

This important turn in Dutch philosophy is reflected in my own development. In my thesis on *Rede en aanschouwing in de wiskunde* (Reason and Intuition in Mathematics, 1935), I made an attempt to reconcile Kantianism with the results of the investigations into the foundations of mathematics. At that time I understood Kant's philosophy in the spirit of Marburg neo-Kantianism as it was then defended in Utrecht, not without essential qualifications, by J. C. Franken (who supervised my thesis), Goedewaagen and Ovink. Like Vredenduin, however, I interpreted the results of the foundational research along the lines of the 'Wiener Kreis'. Already the logicistic trend represented by Carnap, to whose work I owe much, found more response in the Netherlands than Neurath's extreme empiricism. I have occasionally characterized the position which I took at that time as *positivistic criticism.*

In the course of time, however, my horizon broadened and my judgment deepened. After having had the good fortune to be able to attend a course of lectures on the 'Foundations of Mathematics', given in Utrecht by Fraenkel in 1933, my studies in Brussels (1934/35) brought me into contact with Barzin, Errera, Feys, and De Vleeschauwer. At the Paris congress I made the acquaintance of Bernays, with whom I had the privilege to keep up a most interesting correspondence later on; with Heinrich Scholz, whom I visited several times at Münster in 1938; and with Tarski, whom I came to know better during the 'Entretiens d'Amersfoort' (1938). My understanding of intuitionism and of significs benefited by personal contact with Heyting and Mannoury. The correspondence which I kept up with Church as a consulting editor of the *Journal of Symbolic Logic* was also most inspiring.

As a first consequence of this more all-round orientation, I gradually disengaged myself from Kantianism. I did not discover a new philosophical position immediately. As is strongly manifest in my 'Inleiding tot de wijsbegeerte der wiskunde' (Introduction to the Philosophy of Mathematics, 1940), I came to attach much value to scientific objectivity. I continued to reject *geisteswissenschaftliche* methodology and religiously or ideologically determined philosophy. Thus my early positivistic and neo-criticistic leanings were still noticeable.

Naturally, the years of war and occupation also left their mark on my life and thought. I began to study law in 1940 but had to stop in 1942. Towards the end of 1941 I resigned my membership of the 'Genootschap' (which I joined again in 1946), and in 1943 I attended the clandestine meetings of the International Society for Significs. Here I met J. Clay, A. Heyting, B. H. Kazemier, I. Kisch, D. Vuysje and, last but not least, Mrs C. P. C. Fiedler, who was to become my wife in 1947. At this time a new phase in my development began, which lasted until about 1950.

Already at an early age I had been interested in the history of science. Thus as a student I had carefully studied the work of E. J. Dijksterhuis and of my father on the history of mathematics and mechanics. I was encouraged by Pos in my study of the history of philosophy, which I began in 1940, and H. R. Hoetink's lectures on Roman law opened still new perspectives for me. The philosophically relevant conclusions of my historical work may be shortly summarized here.[1]

(1) I observed that Aristotle's conception of science (distinguished by the postulates of deductivity, of evidence and of reality[2]) had, until comparatively recently, strongly influenced the developments of the sciences as well as of systematic philosophy. Modern exact science, however, no longer answers to this conception (the evidence postulate in particular is almost ignored) and accordingly the old ties with traditional systematic philosophy have been severed. This may account for the vehement attacks which philosophers of quite divergent schools have levelled against the latest developments in science.

(2) At the same time, abandoning the evidence postulate meant robbing positivistic criticism of the *geisteswissenschaftliche* method of its foundation. This method could thus be accepted in principle, even though it remained necessary to combat irrationalistic interpretations.

(3) In particular it now became possible to allow for more than one

form of experience. Personally I think that the development of science is determined by at least three forms of experience, viz.: (a) inductive empirical experience, traditionally considered the foundation of natural science; (b) *'understanding'* our fellow-men, which is of fundamental importance for the humanities; this 'understanding' is a primitive datum, it cannot be reduced to our acquaintance with ourselves by way of analogy; (c) the primordial, 'unique' experience by which we are 'reborn'.

(4) The evidence postulate is to be seen as a special case of what I have called *Aristoteles' principle* (Beth, 1946/47, 1952/53). This principle, which is constantly invoked by traditional systematic philosophy, cannot be accepted as universally valid. As a result, traditional metaphysics is seriously undermined.

(5) In spite of this there is still room for ontology as the general theory of reality. Apart from the already mentioned distinction between various forms of experience, a distinction should also be made between various mutually complementary domains of experience (an idea which has also been developed by Bernays[3]), so that this ontology will have to be *pluralistic*.

(6) The acceptance of a pluralistic ontology involves the rejection of the doctrine of the *primacy of consciousness*, which is the basis for all idealistic philosophy (Beth, 1959). Our self-knowledge possesses neither the authentic and primary character nor the degree of certainty that we are inclined to attribute to it. It is at least partly derived from our acquaintance, based on 'understanding', with other persons and with their responses to our behaviour. It is moreover unreliable in so far as it is inculcated upon us with educative intentions. – This view goes back to a conversation with Mannoury.

(7) Finally, I attempted a new interpretation of Kant's classification of judgments into analytic and synthetic judgments, which was subsequently supplemented with a reconsideration of Kant's views on mathematical thought (Beth, 1953/54, 1956/57).

My appointment as a professor in 1946 allowed me henceforth to devote myself exclusively to teaching and to research in those branches of philosophy which were nearest to my heart. It moreover involved the privilege of a regular and close contact with Brouwer and Heyting and with my nearest colleagues Pos and Oldewelt. Among us, philosophers,

Pos was the obvious leader, although I should add with gratitude that his leadership, never of a dominating kind, was always put in the service of others. This gave me from the outset the self-confidence which is so essential to the fulfilment of any responsible task.

As a result of various circumstances, I have had to spend a rather considerable part of my time and energy on practical and organizing work during the years of my professorship. First, of course, since I held a new chair, I had to occupy myself with the organization of classes and of examinations in the subjects entrusted to me. Somewhat later I had to spend much time on the foundation and organization of my own Institute. Shortly after my nomination, moreover, I was involved in the preparation of the Tenth International Congress of Philosophy, for which I acted as a secretary, with Pos as the chairman (*Proceedings*, 1949).

This was soon followed by the foundation (together with Heyting and Van Melsen) of the 'Nederlandse Vereniging voor Logica en Wijsbegeerte der Exacte Wetenschappen' (Dutch Society for Logic and Philosophy of the Exact Sciences) and the extremely laborious work of creating an international organization that was to represent these disciplines. Furthermore I devoted my energies to the 'Algemene Nederlandse Vereniging voor Wijsbegeerte en Psychologie' (Dutch Society for Philosophy and Psychology) and to the editing of its journal, the *Algemeen Nederlands Tijdschrift voor Wijsbegeerte en Psychologie* (Dutch Journal of Philosophy and Psychology). The variety and extensiveness of these activities were due to the already mentioned changed attitude in Dutch philosophy and to the fact that both in the academic world and in philosophical circles the subjects entrusted to me were relatively novel, so that I could not fall back on existing traditions and organizations.

If I have not tried to avoid such tasks, the fulfilment of which was sometimes difficult to reconcile with lecturing and research, and which often failed to give immediate satisfaction, this was due to the consideration that this work should be regarded as extremely useful and even necessary, and that it can be adequately undertaken only by those who are active in the disciplines concerned.

Indeed I should observe at once that these organizational activities, which naturally brought me into contact with many fellow philosophers, certainly did prove to be stimulating for my philosophical work in the long run. This remark leads me to a discussion of the fourth stage of my development, which started in 1950 and is dominated by mathematical logic.

Already at the beginning of my career I had occupied myself with mathematical logic (Beth, 1935b, 1935/36, 1936, 1938), but under the influence of Kantianism, intuitionism and significs I had remained somewhat hesitant towards this subject (Beth, 1937). Once I had overcome my scruples, round about 1939 (cf. Essay VI of the present book), I did not immediately begin to do research in this field. During these years I may have been weighed down by the feeling that I was beyond catching up with mathematical logic. Even in the academic year 1947–48, when in accordance with the terms of my assignment I first included mathematical logic among the subjects of my lectures, I could not bring myself to start with independent research.

The change was brought about by the visit which Alfred Tarski paid to Amsterdam in 1950. After a number of penetrating discussions with him I became confident that there were still many fundamental problems of general interest in mathematical logic which I, with my not very specialized mathematical training, could hope to deal with successfully. Fulbright and Smith-Mundt scholarships enabled me to bring my knowledge more up to date during a visit to Tarski at Berkeley in the spring of 1952. Visits by Kleene, Henkin and Vaught to Amsterdam and a visiting professorship at Johns Hopkins in the spring of 1957 proved to be very stimulating too. Although I did not in the end accept a very attractive permanent position that was offered me in the United States, I retain most valuable memories of both trips to America.

A detailed exposition of the problems which I have tried to deal with in the course of time and of the results which I have reached would be out of place here, but I should like to say at least a few words about them. Even though research in mathematical logic is of a mathematical rather than of a philosophical nature, it cannot be denied that the results often are philosophically significant. Moreover, my choice of the problems to be tackled has always been strongly determined by philosophical considerations. In this connection I should like to make the following observations:

(i) Traditional logic operated on a semantical plane, i.e., in investigating the demonstrative force of arguments it started from the *meanings* of the words and symbols occurring in those arguments. At first modern

101

logic employed a purely *formalistic* method, which left these meanings out of account. Until Tarski gave a proof to the contrary, it was even believed that a semantic approach could never be exact. The equivalence of the formalistic and semantic approaches is guaranteed by Gödel's *completeness theorem* of 1930, which however establishes the connection between the two approaches by a complicated detour (Beth, 1951). The *method of semantic tableaux* which I introduced in 1955 establishes a simple and direct connection and in addition renders Herbrand's and Gentzen's results completely transparent (Beth, 1955, 1962).

(ii) Gödel's completeness theorem belongs to proof theory. A theorem which I have proved about what is called the *method of Padoa* constitutes an analogous result for the theory of definition (Beth, 1953c). In addition, a new way of giving a semantic foundation for a result in the theory of deduction was introduced here. – In this connection I also wish to mention the application of results obtained in the field of logic to the treatment of purely mathematical problems (Beth, 1952a, b, 1953a, b).

(iii) The method of semantic tableaux also provides useful starting-points for a possible automatization of logical reasoning (Beth, 1958a).

(iv) Finally, it should be observed that from the point of view of modern logic the method of semantic tableaux furnishes a better insight into a number of problems and methods of traditional logic. Thus I was able to improve upon Łukasiewicz' interpretation of the *method of exposition* of Aristotle, and to shed new light on the views of Descartes, Locke, Berkeley, Hume, and Kant, with respect to the relation between logical and mathematical reasoning (Beth, 1956/57, 1957c; cf. Łukasiewicz, 1951). – I first heard of the so-called 'Locke-Berkeley problem' during the lectures by Kohnstamm which I had to follow as a prospective school-teacher.

In the study of mathematical logic, three kinds of activity can be distinguished: (a) the development of the system (or systems); (b) the applications of the system, especially in dealing with philosophical or mathematical problems; and (c) the foundation of the system; (c) could be further subdivided into (c_1) the internal foundation of the system, and (c_2) the relation between the system and thought.

The investigations mentioned under (i)–(iv) were mainly concerned with points (b) and (c_1). Point (c_2) will be dealt with in a book that I am preparing now together with Jean Piaget (Beth 1961 [now published]).

My criticism of the recent development of *analytic philosophy* as has taken place in England under the influence of the later Wittgenstein and of John Wisdom and Gilbert Ryle, partly deriving from neo-positivism and from Moore's neo-realism, falls within philosophy in the broader sense (Beth, 1957/58a, b). Obviously, however, my general philosophical views have not developed appreciably in recent years, since most of my time was now devoted to research in the field of mathematical logic. This certainly does not mean that I was no longer interested in general philosophy. On the contrary, I have continued to take part in the activities of the 'Genootschap', and in due time I hope to be able to examine the philosophical consequences of the results reached in mathematical logic. I should also like to contribute to epistemology and to philosophy of science in the future.

In conclusion to these reflections I should like once more to review Dutch philosophy as a whole. The turn from *Weltanschauung* to science which I have already discussed is mirrored in the expansion of the teaching of philosophy at Dutch universities. Whereas there were sixteen (ordinary and extraordinary) chairs in 1940, there are no less than thirty in 1960, while apart from the professors a fairly large number of readers and lecturers is employed. Thus after the heavy losses that we suffered since 1940 as a result of the premature death of men like Franken, Hoogveld, Polak and Pos, there has also been considerable progress.

It is my conviction that if we succeed in putting the available means to effective use by promoting the establishment of adequate links between philosophy and other academic subjects, not only Dutch philosophical life in the more restricted sense but also the pursuit of science in the Netherlands as a whole will profit. The examination system that is now in force promises well for the establishment of such connections, and the objections made against this system by Sassen (Sassen, 1960) are, I think, indeed mistaken. I find the proposals made by the Government Commission rather unrealistic; the existing possibilities of combining subjects ought rather to be extended.

Judging by the present state of affairs, the position of philosophy in higher education has certainly been strengthened. However, the other side of the medal should also be shown. Even though philosophy is in principle independent of religion or ideology, this by no means implies that religion or ideology could do without philosophy. On the contrary, any religious, ideological or other persuasion is dependent on the results of philosophical reflection if it is to attain an intellectually acceptable level.

The mutual relations between philosophy and religion or ideology may be compared with those between mathematics and natural science. Mathematics does not depend on natural science, and it is essential for mathematicians to bear this in mind and to refrain from any appeal to scientific data or methods. If, however, the ties which traditionally link mathematics with the exact sciences were to be completely severed, not only would science be affected, but mathematics, too, would lose an invaluable source of inspiration.

The 'scientism' in philosophy which Pos laments in his article 'Het dal der naoorlogse filosofie' (The Valley of Post-War Philosophy, Pos, 1958) might likewise involve the danger that philosophy will on the one hand fail in its task with respect to the need for a rational persuasion and on the other hand will overlook certain problems which in themselves are philosophically interesting.

There are symptoms indicating that this danger is by no means imaginary. Whereas at the universities the study of philosophy is pursued more actively than in the past, when most philosophical activities took place outside the walls of the university, non-professional or extramural philosophical activities, often focused on matters of religious or ideological persuasion, have strongly diminished both quantitatively and qualitatively. As a result of a more scientific formulation, the ideas of the academic representatives of philosophy meet with little response among people who are above all seeking edification or a persuasion.

Although the complaint has thus, I think, been diagnosed and the etiology has been pointed out, it does not seem so very simple to prescribe an effective therapy. We must not allow philosophy to be given once more the unworthy position of a handmaid of religion or ideology, nor should it descend into the lists and join in the strife of religious and ideological persuasions. Philosophy should rather scan the battle-field of persuasions from the ivory tower of its autonomy and try to determine the underlying philosophical problems that form the real issue, after which these problems can be dealt with using philosophical methods.

Even though this appears to me to be the only correct therapy, I understand quite well that it might seem rather foolish to expect it to bring complete recovery. Does not the individual philosopher have his own persuasion just like other people, so we could argue, and is not it difficult to be impartial when you are a party in the suit?

This legal metaphor, however, current though it may be in this context, does not apply in our case. It suggests the existence of a common legal order which is recognized by all parties, or to which at least all parties are *de facto* subject; such a legal order, however, does not exist here. Thus it is misleading to represent the philosopher as a judge who, by passing his judgment, could settle a dispute.

The question, however, wherein the philosopher's task with respect to the conflict of persuasions consists, already confronts us with a typically philosophical problem. I do not intend to answer this question here (cf., however, Essays II, VIII, and XII of the present book). Suffice it to say for the moment that it is my firm conviction that the philosopher *does* have a task to fulfil here.

REFERENCES

[1] Cf. Beth (1946/47, 1952/53, 1953/54), and Essay V of the present book.
[2] Scholz (1930/31); cf. p. 76 of the present book.
[3] P. Bernays, 'Die Erneuerung der rationalen Aufgabe', in *Proceedings* (1949).

XII

FREEDOM OF OPINION

It is not uncommon that in disputes about scientific or philosophical questions the creed or persuasion of one of the parties in the discussion is referred to, implying that his view-point with regard to the controversial questions is not independent of this persuasion. To ask for the moral justification of this rather common procedure is to call attention to a number of problems in the ethics of scientific discourse, to the discussion of which this essay will be devoted.

Obviously the policy just referred to is quite to the purpose if the disputant in question himself more or less explicitly refers to his persuasion in his argument. Indeed, it might still be acceptable even if he does not make such a reference, provided there is no room for doubt as to the nature of his persuasion.

Matters are quite different if the person in question, no matter for what reason, has not made his persuasion known. If in such a case this persuasion is referred to by outsiders, they attribute a certain persuasion to the person in question on the basis of external data in which he has neither art nor part. All manner of confusion may result from this, as will appear from some quotations concerning myself.[1]

It is gratifying that the wakening interest in more recent trends in philosophy ... does not fail to find an echo in Catholic circles. The publication of this ... introduction to the philosophy of mathematics must surely be considered a most welcome manifestation of this interest, the more so as in treating such an extensive subject the author has taken up a very liberal position. ... "Of course I have an opinion of my own", the author declares in his Preface, " ... but the aim of this book is not to propagandize my own personal views." The author has been as good as his word, as appears above all from his felicitous and successful rendering of the views of non-Catholics by one or two striking phrases and well-chosen quotations.

For example, in sections 44 and 45, dealing with mathematics and mathematical science, where such thinkers as Heisenberg, Bohr, Fraenkel, Frege, Cantor, Hilbert, Beth, Peano, Brouwer, Russell (all, as far as I know, agnostics or Jews) are invoked to suggest that non-Christian science, too, is of great importance.

... in studying the history of the sciences we constantly feel inclined to wish that *Weltanschauung* and natural science would be mixed up less frequently. As a result of the objective of the author's "philosophy of nature" this is no doubt easier than for other views, but is not this very objective the consequence (the more subtle for being unconscious) of a certain *Weltanschauung*? Does the author really believe that ... his own persuasion has not, even if only on the sly, penetrated his book?

The first of the three authors quoted here apparently takes me for a Catholic, whereas the second assumes that I am a Jew or an agnostic. The third reviewer suggests that my philosophical views have been influenced by my *Weltanschauung*, but does not specify it.

Obviously this confusion would not occur if every philosophical author were to state his persuasion explicitly. Indeed this policy is not infrequently recommended, and it is expected that in following it we will be more true to ourselves and that there will be more frankness in scientific discourse.

I propose to devote the remaining part of this essay to the defence of the following theses:

(1) *There is no moral obligation whatsover for a participant in a scientific or philosophical dispute to make his religious or other persuasion known.*

(2) *If a participant wishes to refrain from making his persuasion known, the other parties in the discussion are not morally justified in attributing any persuasion to him.*

Before I start to discuss these theses I should define the term *persuasion* (cf. Essay II). A persuasion is a set of convictions, valuations, attitudes and aspirations, determined by the intellect as well as by the emotions and the will, and which can be expressed to a certain extent in language intelligible to all. Our persuasion is formed in the course of our own lives, partly by environment (upbringing, religion, school, politics), partly by personal experiences (such as professional activities) and, provided a sufficient degree of intellectual maturity has been attained, partly by our own conscious effort.

Whether we like it or not, we all *have* a certain persuasion, and our personal persuasion is part of the personality and thus comparable with our hand-writing, our countenance or our character. Therefore we cannot possibly fall back on somebody else's persuasion; we cannot disown our persuasion without belying our own nature.

Thus no two persons will have *exactly* identical persuasions. However, a large number of people quite often have *more or less* the same persuasion, though this naturally applies to those having a lesser degree of intellectual maturity. In such cases we speak of types of persuasion and of denominational groups. Such a persuasion is to a large extent derived from environment (family, school, church, political party) and is expressed by its exponents without much difficulty in certain stereotyped phrases.

In other cases, however, a persuasion is the fruit of a struggle for insight into one's own personal fate. If we have acquired our persuasion by our own conscious effort, we will be inclined to treasure it as precious personal property and we may not be eager to show this property to others. Moreover, if we should be inclined to make our persuasion known to other people we shall find that suitable means of expression are sometimes not available, and that we have to pay dear for the use of less adequate means of expression.

This consideration would already suffice to explain why not everybody is inclined to make his persuasion known to other people. It would seem to me, however, that the following consideration is of even greater importance. Besides more irenic types of persuasion there are more militant types, besides more tolerant types less forbearing ones, besides types of persuasion which completely govern the behaviour of their followers there are less 'totalitarian' types.

Adherents of a militant persuasion will feel a strong urge to disseminate their views. They will take for granted a similar tendency in other people and perhaps even encourage it. They will not eschew conflicts with more tolerant groups and will be inclined to represent 'proselytism' as a universal moral duty.

Matters are of course quite different for proponents of a more irenic persuasion. Being aware of the fact that arguments about questions of persuasion usually end in violent disagreement, they will be inclined to maintain their aloofness to the utmost. Even for them the time may come when they must no longer be silent. They are not likely, however, to acknowledge a moral obligation to 'proselytize'; they would feel such an obligation rather as interfering with their personal freedom.

Everybody should be free to form his own persuasion and to live in accordance with it. This freedom is fairly commonly recognized. Restrictions of this freedom in certain countries and in certain periods of history, some of which obtain even today, are always felt to be very grievous, but they only concern certain aspects (especially religious, social or political aspects) of persuasions; other, maybe important aspects are hardly interfered with. Indeed you cannot very well forbid or order somebody to be an optimist or a pessimist.

I think the ground has now been prepared for the defence of my two theses. With respect to the first thesis the following line of reasoning seems plausible. Everyone should be free to live in accordance with his own persuasion; for adherents of an irenic persuasion this implies the right not to disclose this persuasion.

This argument is not complete, however. Might not the right to be reticent about one's own persuasion, a right which everybody possesses in principle, turn out to be restricted in the case of somebody who of his own accord engages in a scientific or philosophical discussion? The restrictions would spring from the consideration that if such a discussion is to be successful it is desirable for all participants to make known their persuasions. As was put forward above, this is expected to result in a more candid and outspoken discussion.

The weight of this consideration, however, is extremely dubious. If in a discussion about scientific or philosophical problems the persuasions of the participants are taken into account, this will more probably lead to discord than to mutual candour – as indeed experience teaches us often enough. Moreover, it is by no means an *a priori* certainty that a closer acquaintance with the persuasions of the participants will really contribute to the clarification of their scientific or philosophical view-points; such clarification can be expected only with regard to adherents of a 'totalitarian' persuasion.

Once the consideration just discussed has been eliminated, the argument developed above actually carries weight. In addition I wish to point out once more that many people are, quite understandably and justifiably, reluctant to make public a persuasion which they value as a precious personal property acquired through a hard internal struggle.

I now turn to my second thesis. I wish to defend it by first pointing out that if someone taking part in a dispute exercises his right to maintain silence over his persuasion, it will be extremely difficult for the other parties in the dispute to identify it. Thus the chances are that they will be wide of the mark; the cases quoted from my personal experience may serve to illustrate this. Attributing to somebody a persuasion which is not actually his may obviously have consequences which are highly unpleasant for the person concerned.

Even apart from this, however, to somebody who chooses not to have his private persuasion publicly discussed it is a very unpleasant sensation to have a certain persuasion attributed to him by outsiders. I have first-hand experience in this respect. The tone of the quotations given above was particularly courteous in all three cases and it was in almost flattering terms that I was credited with certain persuasions. Still I must confess that reading these passages at the time made a rather disagreeable impression on me. I should like to analyse this impression a little further.

In the first place I felt that the authors were meddling with my private affairs, and in a rather indelicate way, too. They confined themselves to indicating very sketchily what they thought to be my persuasion and made no attempt to single out those particular elements in this alleged persuasion which could be considered as relevant in the given context.

In the second place I felt placed in a false position. To attribute to an author a persuasion which is not actually his will not only easily cause misunderstanding about the purport of his work but may also damage him in other respects. Therefore, quite naturally, I repeatedly felt the urge to correct the misunderstandings that had arisen. But in what way should I go about to put things right? First I considered the possibility of simply announcing that I was *not* an adherent of the persuasion concerned. But then a rectification in this form did not seem so appropriate, since the express announcement that one does not want to be considered, say, a humanitarian, could easily suggest an anti-humanitarian attitude.

I could only have forestalled this perverse effect by specifying at the same time what I *do* consider as my persuasion. This possibility was in the first place ruled out by the various considerations which I have already mentioned. Besides, such an explanation could give rise to all sorts of misunderstanding, for I have always taken great care not to let my scientific and philosophical work be influenced by my personal persuasion. The above quotations would suggest that I have not been quite unsuccessful in this respect.

My purpose in recording these personal reactions is not to unburden my mind after so many years; after all the experiences described were not all that intolerable. It seemed to me, however, that a short account of these experiences would not be out of place since they are more or less typical for those who object to having their personal persuasion publicly discussed.

In conclusion I should like to supplement and clarify my argument by making a few remarks about the connections between *religion, belief,* and *persuasion.*

Religion is unquestionably among the most important factors in environment influencing the development of our persuasion. This by no means implies, however, that the persuasion is determined by religion.

If somebody professes a certain religious faith, this implies that he wants to belong to a particular religious community and that he accepts the creed of that community. Even so, however, his personal *belief* still depends on the particular ways in which he experiences the religious life within that community and in which he conceives current religious notions. This belief then will become part and parcel of his personal *persuasion,* which will, however, also contain elements from quite different sources.

Although it is thus to be expected that a certain type of persuasion is wide-spread and perhaps even prevalent among the adherents of a certain religion, yet we constantly find that equally convinced believers have strongly divergent persuasions.

Hence, if we attribute to somebody a Buddhistic persuasion, to take a concrete example, this may be misleading in two ways. In the first place the person in question may not be a Buddhist at all. In the second place it may be the case that the person concerned is indeed a true and faithful Buddhist, while adhering to a persuasion which is widely different from the persuasion that may be considered as more or less typical for large groups of Buddhists.

Furthermore, the following observation should be made in this connection. Someone might be called a 'Buddhist' not so much with the intention of identifying him as a member of the religious community of Buddhism as rather to attribute to him a Buddhistic persuasion. In what precedes I have taken it for granted that the first two quotations given above should, *mutatis mutandis,* be understood in this sense, since they are not so much concerned with the religion I profess as with the persuasion I adhere to. However, referring to somebody's persuasion in this way may again give rise to all kinds of misunderstanding.

Let me now reformulate and at the same time summarize the above argument.

If it is thought that somebody's scientific or philosophical views are reducible to his personal persuasion, one risks lapsing into the following line of argument. One knows (or thinks one knows) which religion the person in question professes. It is then concluded that he adheres to 'the' persuasion going with this religion. And finally it is assumed that his scientific or philosophical views are solely determined by his persuasion.

We forget then, however, that somebody's religious convictions do not always unambiguously determine his persuasion. Individual deviations of the type of persuasion that is current within a religious community constantly occur as a result of conscious assimilation of personal experiences. So in the case of scientists or philosophers such deviations are rather to be expected *a priori*. It is hardly conceivable, however, that a persuasion acting as a rigid body of convictions should one-sidedly determine their scientific or philosophical views. More probably, their persuasion will develop in a constant interaction with their scientific or philosophical views.

What was said here about scientists and philosophers of course applies, *mutatis mutandis*, to all who are active in the various spheres of intellectual life. The multiplicity of persuasions is but a reflection of the rich variety and multiformity of life itself.

REFERENCE

[1] In order to keep the discussion as independent as possible from a personal element I purposely refrain from mentioning the names of the authors quoted.

BIBLIOGRAPHY

Abbreviations:

ANTW = *Algemeen Nederlands Tijdschrift voor Wijsbegeerte en Psychologie*
ASI = *Actualités Scientifiques et Industrielles*
PPR = *Philosophy and Phenomenological Research*

A. WORKS BY OTHER AUTHORS

Aebi, M. (1946), *Kants Begründung der 'Deutschen Philosophie'*, Basel.
Bachelard, G. (1949), *La philosophie du non. Essai d'une philosophie du nouvel esprit scientifique*, Paris.
Baeumler, A. (1931), *Nietzsche, der Philosoph und Politiker*, Leipzig.
Bar-Hillel, Y. (1947), 'The Revival of "the Liar"', *PPR* 8.
Becker, O. (1923), 'Beiträge zur phänomenologischen Begründung der Geometrie', *Husserls Jahrbuch* 6.
Bleeker, J. C. (1929), *De beteekenis van de Egyptische godin Ma-a-t* [The Significance of the Egyptian Goddess Ma-a-t], Dissertation, Leiden.
Bohm, D. (1952a, 1952b, 1953), 'Suggested Interpretation of the Quantum Theory in Terms of 'Hidden' Variables', *Physical Review* 85, 87, 89.
Bolland, G. J. P. J. (1889), *De ruimtevoorstellingen* [The Representation of Space], Batavia.
Burkamp, W. (1932), *Logik*, Berlin.
Carnap, R. (1934), *Logische Syntax der Sprache*, Wien.
Corbusier, Le (n. d.), *Le poème électronique*, publisher not indicated.
David, M. (1949), *Les dieux et le destin en Babylonie*, Paris.
Devaux, Ph. (1930), Compte-rendu de: A. N. Whitehead, *Science and the Modern World*, in *Archives Soc. Belge de Philos.* (Bruxelles) 1, fasc. 3.
Diels, H. and Cranz, W. (1934), *Die Fragmente der Vorsokratiker*, 5th ed., Berlin.
Dürr, K. (1938/39), 'Aussagenlogik im Mittelalter', *Erkenntnis* 7.
—— (1951), *The Propositional Logic of Boethius*, Amsterdam.
Dijksterhuis, E. J. (1934), 'De grenzen der Griekse wiskunde' [The Limits of Greek Mathematics], *De Gids* 98.
—— (1956), 'Renaissance en natuurwetenschap' [Renaissance and Natural Science], *Mededelingen van de Koninklijke Nederlandse Akademie van Wetenschappen, afdeling Letteren, Nieuwe Reeks* [Reports of the Royal Dutch Academy, Humanities Department, New Series] 19.
Fraenkel, A. A. and Bar-Hillel, Y. (1958), *Foundations of Set Theory*, Amsterdam.
Frankfort, H. and H. A. (eds.) (1946), *The Intellectual Adventure of Ancient Man*, Chicago.
Freudenthal, H. (1946), *5000 jaren internationale wetenschap* [5000 Years of International Science], Lecture, Groningen-Batavia.

—— (1953), 'Zur Geschichte der vollständigen Induktion', *Archives Int. d'Histoire des Sciences* **22**.

Geyl, P. (1958), *De vitaliteit van de Westerse beschaving* [The Vitality of Western Civilization], Lecture, Groningen.

Gödel, K. (1931), 'Über formal unentscheidbare Sätze der Principia Mathematica und verwandter Systeme I', *Monatshefte für Mathematik und Physik* **38**.

Gonseth, F. (1938), 'Le Congrès Descartes', *Revue Thomiste* **44**.

—— (1939), 'Philosophie mathématique', *ASI* (Paris) **837**.

—— (1948), 'A propos des exposés de MM. Ph. Devaux en E. W. Beth', *Dialectica* **2**.

Grelling, K. (1939), 'Zur Logik der Sollsaetze', *Synthese* **4**.

Grelling, K. and Oppenheim, P. (1938), 'Der Gestaltbegriff im Lichte der neuen Logik', *Erkenntnis* **7**.

Halmos, P. R. (1960), *Naive Set Theory*, Princeton, N. J.

Hartmann, N. (1950), *Philosophie der Natur. Abriss der speziellen Kategorienlehre*, Berlin.

Heinimann, F. (1945), *Nomos und Physis*, Basel.

Hempel, C. G. and Oppenheim, P. (1936), *Der Typusbegriff im Lichte der neuen Logik*, Leiden.

Husserl, E. (1913), *Logische Untersuchungen*, vol. 2, Part 1, 2nd ed., Halle/Saale.

Jaffé, H. L. C. (1956), *De Stijl, 1917–1931. The Dutch Contribution to Modern Art*, Dissertation, Amsterdam.

Jordan, P. (1959), 'Quantenlogik und das kommutative Gesetz', in L. Henkin, P. Suppes and A. Tarski (eds.), *The Axiomatic Method*, Amsterdam.

Kandinsky, W. (1912), *Über das Geistige in der Kunst*, München.

—— (1926), *Punkt und Linie zu Fläche*, Bern-Bümplitz (3rd ed.: 1955).

Kapp, E. (1942), *Greek Foundations of Traditional Logic*, New York.

Kelsen, H. (1941), *Vergeltung und Kausalität*, The Hague. (English translation *Society and Culture*, Chicago 1943.)

Keyser, C. J. (1947), *Mathematics as a Culture Clue and Other Essays*, New York.

Klee, Paul (1945), *Über die moderne Kunst*, Bern-Bümplitz.

Koyré, A. (1946), 'The Liar', *PPR* **6**.

—— (1947a), 'Epiménide le menteur', *ASI* (Paris) **1021**.

—— (1947b), 'Reply', *PPR* **8**.

Kristensen, W. B. (1926), *Het leven uit den dood* [Life from Death], Haarlem.

Lange, F. A. (1877), *Logische Studien*, Iserlohn.

Łukasiewicz, J. (1935), 'Zur Geschichte der Aussagenlogik', *Erkenntnis* **5**.

—— (1951), *Aristotle's Syllogistic from the Standpoint of Modern Formal Logic*, Oxford.

Maritain, J. (n.d.), *La philosophie de la nature*, Paris.

May, Ed. (1937), 'Die Bedeutung der modernen Physik für die Theorie der Erkenntnis', in G. Hermann, Ed. May and Th. Vogel, *Die Bedeutung der modernen Physik für die Theorie der Erkenntnis*, Leipzig.

Melsen, A. G. M. van (1954), *De wijsbegeerte der exacte wetenschap* [The Philosophy of the Exact Sciences], Lecture, Groningen-Djakarta.

Michotte, A. (1946), *La perception de la causalité*, Louvain.

Milhaud, G. (1900), *Les philosophes-géomètres de la Grèce*, Paris.

Mill, J. S. (1843), *A System of Logic, Ratiocinative and Inductive*, London.

Needham, J. (1951), *Human Law and the Laws of Nature in China and the West*, L. T. Hobhouse Memorial Trust Lecture, London.

Neumann, J. von (1932), *Mathematische Grundlagen der Quantenmechanik*, Berlin.

BIBLIOGRAPHY

Newman, J. R. (1956), *The World of Mathematics*, 4 vols., New York.

Panofsky, E. (1955), 'The History of the Theory of Human Proportions', in *Meaning in the Visual Arts*, Garden City, N.Y.

Poincaré, H. (1902), *La science et l'hypothèse*, Paris.

Popper, K. R. (1946), *The Open Society and its Enemies*, 2nd printing, 2 vols., London.

Pos, H.J. (1958), 'Het dal der na-oorlogse filosofie' [The Valley of Post-War Philosophy] in *Keur uit de verspreide geschriften van Dr H. J. Pos* [Selected Papers by H. J. Pos], 2 vols., Arnhem-Assen.

Proceedings (1949) *of the Tenth International Congress of Philosophy (Amsterdam, August 11–18, 1948)* (Library of the Xth Int. Congress of Philosophy, vol. I), Amsterdam.

Russell, B. (1900), *A Critical Exposition of the Philosophy of Leibniz*, Cambridge.

—— (1945), *A History of Western Philosophy*, New York.

—— (1947), *Philosophy and Politics*, London.

Sassen, F. (1960), *Wijsgerig leven in Nederland in de twintigste eeuw* [Philosophical Life in the Netherlands in the 20th Century], 3rd printing, Amsterdam.

Schoenmaekers, M. H. J. (1916) *Beginselen der beeldende wiskunde* [Principles of Plastic Mathematics], Bussum.

Scholz, H. (1921), *Zum 'Untergang' des Abendlandes. Eine Auseinandersetzung mit Oswald Spengler*, 2nd ed., Berlin.

—— (1931), *Geschichte der Logik*, Berlin.

—— (1930/31), 'Die Axiomatik der Alten', *Blätter für Deutsche Philosophie* 4.

Schuppe, W. (1878), *Erkenntnistheoretische Logik*, Bonn.

Selz, O. (1929), 'Essai d'une nouvelle théorie psychologique de l'espace, du temps en de la forme', *Journal de psychologie normale et pathologique* 26.

—— (1930), 'Die psychologische Strukturanalyse des Ortskontinuums und die Grundlagen der Geometrie', *Zeitschrift für Psychologie* 114.

Sigwart, Chr. (1924), *Logik*, 5th ed., with 'Anmerkungen' by H. Maier, Tübingen.

Spengler, O. (1918–22), *Der Untergang des Abendlandes*, 2 vols., München.

Strauss, M. (1938), 'Mathematics as Logical Syntax – A Method to formalize the Language of a Physical Theory', *Erkenntnis* 7.

Tarski, A. (1944), 'The Semantic Conception of Truth and the Foundations of Semantics', *PPR* 4.

— (1956), *Logic, Semantics, Metamathematics. Papers from 1923 to 1938*, Oxford.

Théorie de la preuve (1954), 'Colloque international de logique, Bruxelles 1953', *Revue internationale de philosophie* 8.

Trendelenburg, A. (1862), *Logische Untersuchungen*, 2nd ed., 2 vols., Leipzig.

Ueberweg, F. (1888), *System der Logik*, 3rd ed., Bonn.

Vlastos, G. (1947), 'Equality and Justice in Early Greek Cosmologies', *Class. Philology* 42.

Waerden, B. L. van der (1928), *De strijd om de abstraktie* [The Struggle for Abstraction], Groningen.

Whitehead, A. N. (1926), *Science and the Modern World*, Cambridge.

—— (1929), *Process and Reality*, Cambridge.

—— (1933), *Adventures of Ideas*, Cambridge.

Worringer, W. (1908), *Abstraktion und Einfühlung*, München (6th ed.: 1918).

SCIENCE A ROAD TO WISDOM

B. WORKS BY E. W. BETH

1935a. *Rede en aanschouwing in de wiskunde* [Reason and Intuition in Mathematics], Dissertation, Groningen.
1935b. 'Sur un théorème concernant le principe du tiers exclu', *2me Congrès National des Sciences* (Bruxelles, 19–23 juin 1935).
1935/36. 'La métamathématique et ses applications au problème de la non-contradiction de la logique et de l'arithmétique', *Christiaen Huygens* **14**.
1936. 'Démonstration d'un théorème concernant le principe du tiers exclu', *Bulletin de l'Académie Royale de Belgique* (Classe des sciences), séance du 5 mai 1936.
1937. 'L'évidence intuitive dans les mathématiques modernes', *Travaux du IXe Congrès international de philosophie (Paris, 1–6 août 1937)*, 6.
1938. 'Une démonstration de la non-contradiction de la logique des types', *Nieuw Archief voor Wiskunde* **19**.
1940. *Inleiding tot de wijsbegeerte der wiskunde* [Introduction to the Philosophy of Mathematics], Antwerpen-Brussel-Nijmegen-Utrecht (2nd printing: 1942).
1946/47. 'Historical Studies in Traditional Philosophy', *Synthese* **5**.
1951. 'A Topological Proof of the Theorem of Löwenheim-Skolem-Gödel', *Indag. Math.* **13**.
1952a. 'Existence of Complete Models for Extensions of the First-order Predicate Calculus', *Bull. Amer. Math. Soc.* **58**, Abstract 501.
1952b. 'Observations métamathématiques sur les structures simplement ordonnées', *Applications scientifiques de la logique mathématique, Actes du 2e colloque international de logique mathématique* (Paris, 25–30 août 1952).
1952/53. 'The Prehistory of Research into Foundations', *British J. Philos. of Sc.* **3**.
1953a. 'Sur la description de certains modèles d'un système formel', *Actes du XIe Congrès international de philosophie (Bruxelles, 20–26 août 1953)*, vol. V.
1953b. 'Some Consequences of the Theorem of Löwenheim-Skolem-Gödel-Malcev', *Indag. Math.* **15**.
1953c. 'On Padoa's Method in the Theory of Definition', *Indag. Math.* **15**.
1953/54. 'Kants Einteilung der Urteile in analytische und synthetische', *ANTW* **46**.
1955. 'Semantic Entailment and Formal Derivability', *Mededelingen Koninklijke Nederlandse Akademie van Wetenschappen, Nieuwe Reeks*, **18**.
1956/57. 'Über Lockes "Allgemeines Dreieck"', *Kantstudien* **48**.
1957a. 'Le savoir déductif dans la penseé cartésienne', in *Descartes*, Paris.
1957b. *La crise de la raison et la logique* (Collection de logique mathématique, Série A), Paris-Louvain.
1957/58a. 'Geformaliseerde talen en normaal taalgebruik' [Formalized Languages and Ordinary Usage], *ANTW* **50**.
1957/58b. *Naschrift* (Postscript), *ibid.*
1958a. 'On Machines which Prove Theorems', *Simon Stevin* **32**.
1958b. '"Cogito ergo sum" – Raisonnement ou intuition?', *Dialectica* **12**.
1959. *The Foundations of Mathematics (Studies in Logic)*, Amsterdam.
1961. (with J. Piaget) *Epistémologie mathématique et psychologie. Essai sur les relations entre la logique formelle et la pensée réelle*, Paris.
1962. *Formal Methods. An Introduction to Symbolic Logic and to the Study of Effective Operations in Arithmetic and Logic*, Dordrecht.

INDEX OF NAMES

117

118

INDEX OF SUBJECTS